Cognitive Systems Engineering for User-Computer Interface Design, Prototyping, and Evaluation

Cognitive Systems Engineering for User-Computer Interface Design, Prototyping, and Evaluation

Stephen Andriole
Drexel University

Leonard Adelman
George Mason University

LEA LAWRENCE ERLBAUM ASSOCIATES, PUBLISHERS
1995 HILLSDALE, NEW JERSEY   HOVE, UK

Lawrence Erlbaum Associates, Inc., Publishers
365 Broadway
Hillsdale, New Jersey 07642

Library of Congress Cataloging-in-Publication Data

Andriole, Stephen J.
Cognitive systems engineering for user-computer interface design,
prototyping, and evaluation / Stephen Andriole, Leonard Adelman.
p.   cm.
Includes bibliographical references and index.
ISBN 0-8058-1244-X
1. User interfaces (Computer systems).   2. Systems engineering.
I. Adelman, Leonard, 1949–   . II. Title.
QA76.9.U83A53   1995
004′.01′9—dc20   94-30087
                                                                              CIP

Books published by Lawrence Erlbaum Associates are printed on acid-free
paper, and their bindings are chosen for strength and durability.

10 9 8 7 6 5 4 3 2 1

To our wives and kids—

Denise, Katherine, and Emily Andriole
Lisa, David, and Lauren Adelman,
for their love, understanding, patience, support,
and willingness to let Dad get back to work . .

Steve Andriole, Bryn Mawr, Pennsylvania
Len Adelman, Herndon, Virginia

Contents

Preface

Many of us have used computer programs that were difficult if not impossible to fathom. Some of us have been active in the design of user-computer interfaces and interaction routines designed to make these systems much easier to use. Many designers of interfaces and interaction routines have relied on the impressive findings of the human factors and ergonomics research community; others have begun to enhance these results with emerging results from cognitive science.

This book is the result of successes and failures, and our casting "the analytical net" as far as we could throw it. We have relied upon conventional human factors and ergonomics, psychology, cognitive science, and even systems engineering to design, prototype, and evaluate user-computer interfaces and interaction routines. We have applied the "methodology" described here to a variety of domains. The book is as much about this methodology as it is about the applications.

It is difficult to separate user-computer interface and interaction routine design, prototyping, and evaluation from the emerging information technology available to all systems engineers. Sometimes it is very hard to tell which interface feature was inspired by the technology or a bona fide user requirement. This issue will remain with us; "requirements pull" versus "technology push" is a very real issue for UCI designers.

There is also the need to remain disciplined and systematic. We wanted to communicate a particular design/prototyping/evaluation "discipline"; that is, a set of steps that when taken together describe a methodology likely to result in a better interface.

But what is "better"? We have tried to measure performance qualitatively and quantitatively. An interface that makes users feel good or makes them believe that they are more productive than they were is only better if in fact actual performance improves. We have—perhaps like you—observed the phenomenon where users "like" interfaces that degrade their performance—and vice versa. Evidence reduces the likelihood of this effect.

Motivation for the book came from a desire to integrate and synthesize findings from several fields and disciplines into a methodological process that is evidence-based. We describe the process and then illustrate how it has been applied to several case studies. We also wish to demonstrate the interdependencies among requirements, information technology, design, and evaluation methodology that it is impossible to design a new or enhance an old interface without reference to requirements, available information technology, and a disciplined methodological process that supports evidence-based design and evaluation.

Finally, we regard user-computer interface and interaction routine design, prototyping and evaluation as another activity that falls comfortably under the overarching systems engineering "umbrella." This, we believe, is an important concept. We believe that rather than continue to treat user-computer interaction, human-computer interaction, or computer-human interaction as special fields of inquiry, it is time to regard it as a systems engineering process. "Cognitive systems engineering" is just another application of the generic systems engineering process.

—Stephen Andriole
—Leonard Adelman

Acknowledgments

We would like to thank our respective universities—Drexel University and George Mason University—for their direct and indirect support of the research described in this book.

As noted in the book, the case studies were supported by a variety of organizations and individuals—to whom we owe thanks.

We would like to thank our families for their usual—and now legendary—patience and support.

We would also like to thank our friends and colleagues for listening to our papers at conferences and informal meetings, and our students for challenging us and filling in the gaps of our thinking at—it seems—the most critical times.

—Steve Andriole
—Len Adelman

Cognitive Systems Engineering in Perspective

This book seeks to answer the question: Can findings from cognitive science enhance the user–computer interaction process? The book recognizes that user–computer interfaces (UCIs) are often essential parts of an information or decision support system; in fact, UCIs are often critical components of software-intensive systems of all kinds. But we recognize from the outset that the design, prototyping, and evaluation of user–computer interfaces are part of larger systems and are therefore ideally designed, developed, and evaluated as part of a larger design and development process—or *life cycle*.

This book thus describes the process by which requirements—be they functional, nonfunctional, or display oriented—are converted first into prototypes and then into working systems.

Although the process may at times seem almost mysterious, there is in fact a methodology that drives the process, a methodology that is defined into terms of an adaptive life cycle.

There are a number of steps or phases that comprise the standard life cycle; there are methods, tools, and techniques that permit each step to be taken. This book describes our effort to implement this process to enhance user–computer interaction. More specifically, the book describes a methodological approach that seeks to identify and apply findings from cognitive science to the design, prototyping, and evaluation of user–computer interfaces.

THE DESIGN AND DEVELOPMENT BACKDROP:
INFORMATION AND DECISION SYSTEMS ENGINEERING

The steps that constitute a generic design and development life cycle follow. They suggest the primary tasks that must be taken if requirements are to evolve into working system concepts. As is evident in the steps, there are objectives (what you want the step to yield), methods (methods, models, approaches and concepts available to you to achieve the objective), and tools (off-the-shelf [OTS] software systems) that have incarnated the methods. We have highlighted in *italic* the steps that pertain to the cognitive engineering of user–computer interfaces and the user–computer interaction process.

1. *Analyze Requirements*
   a. Method(s): *Interview* and observe
   b. Tool(s): OTS Software: Outline/idea processors
   c. Objective: *First cut at user requirements*; Document for continued use
2. *Model Requirements*
   a. Method(s): Hierarchical decomposition, IPO, Causal, *Cognitive maps/Mental models*, Simulations
   b. Tool(s): OTS software: Hierarchical tools, Simulation tools, General purpose modeling tools
   c. Objective: *Initial transition from observational and textual data and information into structured, organized model of user requirements*
3. *Prototype Requirements*
   a. Method(s): Throwaway/Evolutionary: *Screens, Interactions, Output (External behavior of system)*
   b. Tool(s): OTS Software: *Screen formatters, Interactive prototypers*, Code generators, UIMSs
   c. Objective: Throwaway Prototype: *Convert requirements model into a system concept that demonstrates UCI and Overall functionality*; Evolutionary Prototype: Begin transition to full-scale production (*Build a little, Test a little*)
4. *Specify Software Requirements*
   a. Method(s): DFDs, ERDs, NSs, GS, DeMarcos
   b. Tool(s): OTS Software: Conventional CASE tools
   c. Objective: Represent the software design via consistent notation to "guide" production of code and document the conversion process
5. *Develop Software*
   a. Method(s): Structured programming, programming "conventions"

    b. Tool(s): Programmer workbenches; CASE tools; Programming "environments" (e.g., Ada)

    c. Objective: Software that works

6. *Test software*
7. *Implement system*
8. *Evaluate (Prototype) system*
9. *Modify system*
10. *Document system*
11. *"Maintain" system*
12. *Advance to Step 1 . . .*

This life cycle is derived from classic systems engineering life cycles and life-cycle models that pertain to the design and development of software-intensive systems (Andriole & Freeman, 1993; Blanchard, 1990; Eisner, 1989; Sage, 1992).

THE PRIMACY OF REQUIREMENTS, PROTOTYPING, AND EVALUATION

The detailed prototyping (and prototype evaluation) process is as follows:

- From requirements, which in this case are in the form of tasks, scenario-based descriptions, and cognitive maps, derive functional profiles of what the system already does and might do via enhancements.
- Identify the requisite knowledge from cognitive science (expressed in a set of "findings" for which there are empirical results), human factors, and the display and interaction technology (color, graphics, icons, multimedia, and the like) that will make the prototype as responsive as possible to the requirements.
- Distill the synthesized options down to the most promising. This step involves making a series of informed decisions about the kinds of displays and interaction routines to be prototyped that will have the greatest impact on human performance. Here we try to determine if cognitive science can aid the design and development of displays and user computer interaction routines that will enhance human performance.
- Develop a set of screen displays—*storyboards*—that represent, in effect, hypotheses about what kinds of displays are most likely to enhance performance. Develop the displays via a flexible, adaptive tool, a tool whose output can be imported into any number of applications programs.

- Connect the screens into mini sequences that each represent an aggre-gate hypothesis about, for example, how to enhance the specific tasks or subtasks.

- Implement the screen displays and sequences for subsequent testing and evaluation.

- Develop a testing and evaluation (T&E) plan consisting of details about subjects, hypotheses, controls, data collection, measures of effective-ness, and the like.

- Pretest to whatever extent is necessary, feasible, and practical.

- Run the experiments to determine the impact on performance.

- Summarize the results and interpret the findings vis-à-vis the specific hypotheses tested as well as the potential generalizability of the findings.

COGNITIVE SCIENCE AND REQUIREMENTS MODELING

The process can also be understood in terms of a set of bases or *bins* from which we extract useful and pertinent findings and opportunities. For exam-ple, we look at the requirements (according to the process described earlier) and reach into the requirements data and extract those that we believe represent high-leverage tasks. We look at the findings from cognitive science and extract those that we feel can impact the requirements; and will look at information technology—broadly defined—to determine how we can imple-ment the ideas in displays and interaction routines. This last step is challenging because we must look at available technology that can be implemented on the target systems (not just on our own particular wish lists).

There are four major case studies presented in this book. The first is *conceptual.* It addresses a domain—strategic air defense and intelligence—that is intensely user–computer-interaction intensive and proposes a set of displays and interaction routines that may enhance performance. The second case study deals with the interface of the patriot missile defense system, the third with military air traffic control, whereas the fourth examines the inter-face surrounding decisions to launch torpedos in an antisubmarine context.

A set of generic requirements—and some high-level findings from cog-nitive science—help describe the process by which we engaged in cognitive systems engineering.

Some examples of generic inference and decision-making requirements are presented next. However, note that this short list is followed by a list of display requirements, because nearly all substantive requirements have display correlates. (We subsequently made the generic functional/substan-tive and display requirements particular to the case studies described in the book.)

Some requirements appear below, followed by generic display requirements:

Substantive Inference-Making Requirements:
- Assessment of events and conditions; inferences about likely event outcomes
- Identification of high-probability crisis situations
- Mechanisms for discriminating between "signals" and "noise"
- Techniques for minimizing "false alarms" and maximizing "hits"
- Inter- and intragroup communications
- Heuristics for adaptive monitoring; methods for quasi-automated monitoring and warning

Inference-Oriented Display Requirements:
- Distilled displays of threat
- Displays of geographical vulnerability
- Historical displays of comparative cases
- "What-if" displays

Substantive Decision-Making Requirements:
- Tactical and strategic option generation
- Option evaluation
- Assessment of operational constraints
- Information/knowledge processing
- Search for historical precedents (successful and unsuccessful)
- "What-if" analyses
- Calculation of decision option outcomes
- Inter- and intragroup communications

Decision-Making-Oriented Display Requirements:
- Displays of costs and benefits
- Displays of alternative options
- Displays of option rank orders
- Displays of communications networks
- Displays of decision option outcomes
- Methods and techniques for rapid information/knowledge processing

These and other functions and tasks are representative of the kinds of activities that define system and user–computer interface interaction (UCI) requirements. The challenge remains as to how to design and develop interactive systems that will satisfy these requirements, while simultaneously minimizing operator workload. One way to accomplish these goals is through cognitive systems engineering.

Presented next is a set of findings from cognitive science representative of the kinds of results that can be obtained via empirical research. The findings are organized around the two exemplar generic domains of inference and decision making.

Inference-Making Findings:
- Inference-making performance is task dependent
- Human information processors (HIPS) have poor insight into their own inferential processes
- Cognitive feedback can significantly improve task performance
- HIPS are susceptible to non-linear-based deception
- HIPS use simple and complex "causal" schemas to explain behavior and make inferences
- HIPS often reason from experiences, analogies, or cases
- HIPS use plausible scenario generation to test and generate hypotheses
- HIPS rely on cues-to-causality to explain current and future behavior
- HIPS process less information than they think they use to make inferences
- Experts are as susceptible to poor insight about their own inferential processes as novices
- HIPS are unsure about how to optimize data, information, and knowledge in inference making
- HIPS are prone to cognitive reductionism to reduce workload and overall mental effort
- The perception of information is not comprehensive, but selective
- HIPS tend to underemphasize base-rate information
- HIPS weigh the importance of information on the basis of its perceived causal meaningfulness not based on its statistical diagnosticity
- HIPS are susceptible to the "availability bias" or to the tendency to recall recent or highly publicized events
- HIPS selectively perceive data, information, and knowledge on the basis of experience
- HIPS are susceptible to confirmation biases that filter incoming information according to preconceived ideas and views
- The way information is presented often determines how it is perceived
- HIPS tend to be conservative inference makers

Decision-Making Findings:

- Human decision makers (HUDMS) tend to simplify decision problems by setting outcome aspirations
- HUDMS often choose the first decision alternative that satisfies the outcome aspiration(s)
- HUDMS often simplify decision problems by only considering a small number of alternatives
- HUDMS use analogies to generate and compare options
- HUDMS weigh criteria to rank order decision alternatives
- HUDMS selectively perceive data, information, and knowledge; focus on confirming (versus disconfirming) information; and tend to anchor their judgments
- HUDMS tend to attribute decision outcomes to chance or to the complexity of the problem (not to their own decision-making deficiencies)

This short list of findings suggests how cognitive science can inform the systems engineering process. User-computer interaction (UCI) routines, displays, dialogue, and other communications/presentation techniques can be used to narrow the gap between the way humans make inferences and decisions and the way systems support them. When combined with the power of advanced information technology, new system concepts can be developed and tested.

THE EMERGING ROLE OF ADVANCED
INFORMATION TECHNOLOGY

Substantive and display requirements and findings from cognitive science beg the technology challenge. Can advanced technology help? How? Where? What kinds of technologies? How do we know that they will work? The working premise here is that there are a variety of advanced (and not-so-advanced) information technologies that can greatly aid the information manager and inference and decision maker.

There are a variety of technological solutions available to the systems engineer. Some are listed next. These (and other) ideas need to be tested in the context of some specific scenarios as working prototypes before they become elements of command center capabilities or operational procedure. The list contains many ideas that simply could not have been proposed just 5 years ago; 10 ago they would have been judged unrealistic or prohibitively expensive. It is essential that we exploit advanced information technology; at the same time, it is essential that our technological solutions be requirements driven (and not unduly "pushed" by the technologies themselves) and informed by what we know about human information processing, decision making, and problem solving:

- Adaptive real-time/online message processing/routing systems
- Advanced message typography systems
- Analogical (case-based) data/knowledge-based analyses
- Spatial/iconic-based search techniques
- Hypertext-based ultra-rapid reading and information/knowledge processing
- Multimedia (text, photographic, sound, video, etc.)-based situation displays
- Online historical data/knowledge bases of previous, pertinent planning and decision-making episodes
- Direct manipulation graphic displays for "what-if" analyses and other functions
- Knowledge-based option generators
- Real-time simulation/animation of decision option outcomes
- Shared data/knowledge bases for group decision making, among other ideas

These ideas—among many others—are, in principle, viable solutions to complex analytical problems. Are they any good? Will they enhance human performance? How can they be married to cognitive science to yield the greatest leverage?

THE EMERGING FIELD OF COGNITIVE SYSTEMS ENGINEERING

Recently a number of applied psychologists, information scientists, and systems engineers have joined forces to create a new field of inquiry currently known as *cognitive systems engineering*. The key word in the label is *engineering*, because it reflects an intention to *apply* what we know about human cognitive information processing to larger systems engineering efforts. Had *cognitive engineers* been around (in name, at least) in the mid-1970s, they would have *invented* windows, the mouse, navigational aids, and interactive graphic displays. Today's cognitive engineers anchor their design recommendations in what we know about human information processing; what we "know" is ideally determined by theoretically inspired experimentation that yields empirically verifiable results.

It is extremely important to note at the outset that cognitive science should be synthesized with human factors research (and information technology) to optimize the likelihood of designing an effective interface and interaction routines likely to enhance performance.

Our work in cognitive systems engineering has been oriented to demonstration, that is, to demonstrate that it is in fact possible to leverage findings from the cognitive sciences synergistically with advanced information technology to enhance human performance. More specifically, we have attempted to generate the following:

- Empirically verified findings about how displays might be modified to enhance human–computer performance
- A methodology for conducting retrofits of existing systems or analysis and tests of planned or evolving systems.

THE ORGANIZATION OF THE BOOK

The book is also a kind of "demonstration project." It describes a process by which cognitive science can be used to design new—or retrofit old—user–computer interfaces. It does so via a set of case studies that have been conducted over the past few years.

This first chapter places cognitive engineering within the larger context of systems engineering. It argues for serious and extensive requirements analysis prior to UCI design, and it argues for prototyping to determine the effect of alternative UCI designs. Chapter 2 surveys the broad field of cognitive science and the findings that have potential relevance to UCI design and enhanced UCI performance. Chapter 3 looks at information technology and how it can be married to findings from cognitive science (along with the more traditional human factors findings) to design better user–computer interfaces and interaction routines. Chapter 4 positions and describes the case studies in the context of the field. Chapters 5, 6, 7, and 8 present the case studies. Chapter 9 develops some guidelines for the synthesis of cognitive science, human factors, and information technology to satisfy UCI requirements.

The Cognitive Bases
of Design

Cognitive information systems engineering is a relatively new field of inquiry that calls for the design and development of computer-based information systems consistent with what we know about how humans process information and make decisions. To quote Norman (1986, p. 31), cognitive engineering is "a type of applied Cognitive Science, trying to apply what is known from science to the design and construction of machines." It represents the attempt to base systems design on cognitive research findings regarding how people think; to aid—not replace—the human problem solver; to design systems that utilize human strengths and compensate for our cognitive limitations; and to more systematically integrate display formats and analytical methods with intuitive thought. Put quite simply, the goal of cognitive engineering is to have human cognitive processes, not information technology, drive the human–computer interaction process.

This chapter overviews the findings from a number of different areas of cognitive research that are potentially applicable to the cognitive engineering of advanced information systems. In contrast to many important cognitive research efforts, we do not focus on the cognitive engineering of the user interface for the sole purpose of better using a software package, such as a text editor (Card, Moran, & Newell, 1983), a word processor (Norman, 1986) or a spreadsheet (Olson & Nilson, 1987). Instead, our focus is on enhancing organizational problem solving, that is, on designing the user interface to advanced information systems so that they improve our ability to perform the complex inference and decision-making tasks found in organizational settings (Andriole & Adelman, 1991; Wolf, Klein, & Thordsen, 1991). To quote Woods and Roth (1988, p. 417), "Cognitive engineering is about human behavior in complex worlds."

The sections here review the research literature for four cognitive process areas: judgment, decision making, memory, and attention, in which the latter includes reference to work-load research. The review is not meant to be exhaustive, but, rather, selective. In particular, the majority of the review focuses on judgment and decision processes, for they represent the superordinate processes. Stated somewhat differently, judgment and decision processes represent the higher order cognitive skills that utilize memory and attention skills for effective problem solving under high work-load conditions. This is not meant to imply that memory and attention processes are not important. Obviously, one cannot make good judgments and decisions without remembering critical information and accurate relations between events and objects. Nor can one make good judgments and decisions without attending to critical aspects of the environment. Rather, our principal focus is on the higher order cognitive processes they support because (a) we think these superordinate processes will be the critical focus of many future cognitive engineering efforts, and (b) they have been relatively neglected in many efforts thus far.

In closing this introductory section, we note that for each cognitive research area we briefly overview the types of tasks used in performing the research. Although we realize that the reader's primary concern is learning about the findings of different areas of research, there is substantial empirical research (e.g., Hammond, Stewart, Brehmer, & Steinmann, 1975; Hogarth, 1981; Payne, 1982) demonstrating that human information processing and decision making are extremely sensitive to task characteristics. Carroll (1989, p. 4) has even gone so far as to question the viability of a cognitive engineering discipline because "the science base in which design deductions must be anchored is too general and too shallow vis-a-vis specific contexts and situations in the world."

Although we disagree with Carroll, we caution readers against blindly generalizing basic research findings to complex organizational settings. This review is but a part of the larger project. That project is directed toward (a) identifying display technology that, on the basis of cognitive research findings and the substantive requirements of the task, should enhance cognitive processing; (b) developing new displays consistent with this cognitive engineering perspective; and (c) performing empirical evaluations to determine whether performance is indeed improved in complex problem solving. We focus on the different tasks used in basic research as a means of more effectively moving toward that end.

Finally, we use the phrase "cognitive information systems engineering" synonymously with other phrases in the literature, such as "cognitive systems engineering" (Hollnagel & Woods, 1983), "cognitive engineering" (Rasmussen, 1986; Woods & Roth, 1988), and "cognitive ergonomics" (Falson, 1990).

JUDGMENT AND DECISION MAKING

We begin by considering various paradigms in the literature for considering judgment and decision making. The point in doing so is to illustrate the tight coupling between these two cognitive processes. Then we consider the research for each process in turn. Simon (1960) has used three categories to describe decision-making activities: intelligence, design, and choice. *Intelligence* refers to the activities inherent in problem identification, definition, and diagnosis. It is, as Huber (1980) pointed out, the conscious process of trying to explore the problem in an effort to find out the current state of affairs and why it does not match our desires. *Design* refers to those activities inherent in generating alternative solutions or options to solving the problem. It involves "identifying items or actions that could reduce or eliminate the difference between the actual situation and the desired situation" (p. 15). *Choice* refers to those activities inherent in evaluating and selecting from the alternatives. It is the action that most people think of when one makes a decision.

As Huber (1980) and others (e.g., Andriole, 1989a; Sage, 1986; Wohl, 1981) have pointed out, decision-making activities are a subset of problem-solving activities. For example, the first three steps in Huber's 5-step problem-solving paradigm are those activities that require (a) problem identification, definition, and diagnosis; (b) the generation of alternative solutions; and (c) evaluation and choice among alternatives. These steps are conceptually identifical to Simon's decision-making categories. The fourth step in Huber's paradigm involves activities inherent in implementing the chosen alternative. The fifth step involves activities inherent in monitoring the implemented action in an effort "to see that what actually happens is what was intended to happen" (Huber, 1980, p. 19). If there is a significant mismatch between the actual and desired state of affairs, one returns to the first step—exploring the problem.

Wohl (1981) presented a paradigm within the context of military tactical decision making that expands on the activities in Simon's and Huber's paradigms. The anatomy of Wohl's SHOR (Stimulus–Hypothesis–Option–Response) paradigm is presented in Figure 2.1. Intelligence activities are differentiated between the Stimulus and Hypothesis elements of the SHOR paradigm. In particular, the Stimulus element is comprised of data collection, correlation, aggregation, and recall activities; it naturally includes many of the activities also included in Huber's last problem-solving stage, that of monitoring the situation.

GENERIC ELEMENTS	FUNCTIONS REQUIRED
Stimulus (Data) **S**	**Gather/Detect**
	Filter/Correlate
	Aggregate/Display
	Store/Recall
Hypothesis (Perception Alternatives) **H**	**Create**
	Evaluate
	Select
Option (Response Alternatives) **O**	**Create**
	Evaluate
	Select
Response (Action) **R**	**Plan**
	Organize
	Execute

FIG. 2.1. The SHOR Paradigm.

The Hypothesis element is that aspect of intelligence that involves creating alternative hypotheses to explain the possible cause(s) of the problem, evaluating the adequacy of each hypothesis, and selecting one or more hypotheses as the most likely cause(s) of the data. It is important to note that more than one hypothesis can be appropriately selected either because of the uncertainty and/or ambiguity in the data (Daft & Lengel, 1986), or because there is more than one cause of the problem (Hammond, 1966). Regardless, on the basis of the selected hypothesis (or hypotheses), the decision maker (and senior associates) generate options for solving the problem. The Option element explicitly differentiates between option creation, evaluation, and selection activities. Finally, on the basis of the selected option, the decision maker (or decision-making team) takes action, which includes the planning, organization, and execution of a Response to the problem, analogous to the fourth step in Huber's problem-solving framework. (Note: Wohl's distinction between the Hypothesis and Option components is analogous to Hogarth's [1987] general distinction between prediction and evaluation.)

The Option component follows the Hypothesis component in all three paradigms. Options are not generated in a vacuum. Rather, they are generated in response to our hypotheses with regard to what is happening to us, why it is happening, and its implications. Now this does not imply that every option generation situation requires a causal focus. For example, when one breaks one's sunglasses, as one of the authors did, one does not need to consider why they were broken to generate options regarding where to get a new pair. (However, one would need to consider why they were broken to generate options for not breaking them again.) Rather, the paradigms imply that in many situations a causal focus is essential to good option generation and decision making. More generally, judgments about the situation affect one's decisions about what to do.

There is considerable empirical support for the premise that judgments about the situation affect option generation and evaluation. One source, for example, is the problem-solving research using protocol analysis. To quote Newell and Simon (1972, p. 783), "This pattern of symptom-remedy has by now become familiar to us in a variety of contexts. It is means-ends analysis, for the moves generated are relevant to 'remedying' the feature that serves as symptom." We now briefly consider the task characteristics of problem-solving research and then important findings.

The typical procedural approach in problem-solving research is to ask people to talk out loud as they attempt to solve a provided problem. Numerous different types of problems have been studied (e.g., chess, programming, physics, logic, cryptarithmetic) as well as alternative representations of the same problem structure, called *problem isomorphs* by Simon and Hayes (1976). The reason for having people talk out loud is to trace their problem-solving behavior as it occurs; thus, it is not susceptible to logical fallacies in recall or in the hindsight bias (e.g., see Ericsson & Simon, 1984). The audiotapes (or protocols) are transcribed for subsequent analysis, which is why the approach is often referred to as protocol analysis. The typical analysis approach has been to infer decision processes by constructing problem behavior graphs (Newell & Simon, 1972), which resemble flow charts composed of nodes (knowledge states) connected by arrows (process operations). However, other methods have been used too.

Problem-solving research is vast and has generated numerous findings. Of importance here, however, is the identified strong relationship between the Hypothesis and Option components in Wohl's (1981) paradigm. For example, in their review and discussion of de Groot's (1965, 1966) process-tracing analysis of novice and expert chess players, Newell and Simon (1972, pp. 775, 776) pointed out that:

> during the first moments—for example, 15 seconds more or less—during which he is exposed to a new position, a skilled human player does not appear to engage in a search of move sequences. Instead, he appears to be occupied with perceiving the essential properties of the position ... which will suggest possible moves to him and help him to anticipate the conse-

quences. He appears to be gathering information about the problem, rather than seeking an actual solution.

In short, the human problem solver is testing hypotheses to explain the data or "defining the situation" (p. 761). We have chosen to use the word *judgment* instead of *cause* to convey a broader focus on all aspects of the situation that may affect one's decision making.

Once he or she has done so, possible solutions are generated and evaluated, apparently in depth. Again, to quote Newell and Simon (1972, p. 751), "Humans playing chess spend much of their time searching in the game tree for the consequences of the moves they are considering. . . . The search is highly selective, attending to only a few of the multitude of possible consequences." Klein (1989) found the same results with urban and wildlife fire ground commanders and design engineers. People do not typically search for a complete list of options, but rather for a quality one based on his or her assessment of the situation and goals.

From the perspective of previous option generation research, what is particularly interesting about the problem-solving research reviewed by Newell and Simon (1972) and Ericsson and Simon (1984), and replicated by Klein (1989) outside the laboratory, is the finding that expertise is related to a person's ability to generate correct problem representations, that is, the correct hypotheses or causes for the observed data. To stay with the chess example, de Groot (1966) demonstrated that chess grandmasters were significantly better able than good but weaker players to reproduce a chess board after an extremely brief exposure (3–7 sec). However, this capability was totally dependent on the meaningfulness of the chess board; the superiority vanished if the chess pieces were randomly distributed on the board. Chase and Simon (1973) replicated and extended this finding. Better players are better able to infer causes (i.e., possible opponent strategies), but only for meaningful data.

A number of studies have shown a relation between Hypothesis and Option elements of the SHOR model. For example, Adelman, Gualtieri, and Stanford (in press) found that the causal focus of the information presented to people affected the type of options they generated and selected. Bouwman (1984, p. 326), who analyzed the protocols for three Certified Public Accountants and five graduate students evaluating financial cases, found that "experts regularly summarize the results, and formulate hypotheses. Such 'reasoning' phases further direct the decision making process." A longitudinal, process-tracing study by Schweiger, Anderson, and Locke (1985), which used the UCLA Executive Decision Game as the task, found that subjects who engaged in causal reasoning performed significantly better than those who did not. And Cohen's (1987) study of Air Force pilots, found that:

> the pilot who adopts a worst case strategy is not really suppressing uncertainty; he knows perfectly well that other outcomes are possible. Rather, he is adopt-

ing assumptions which enable him to focus on a concrete, causally modeled state of affairs as opposed to an abstract, non-realizable average or expected value. He may subsequently wish to undo these particular assumptions and explore another set, which implies another concrete, causally modeled state of affairs. (p. 36)

Research by Klein and his associates (e.g., Klein, 1989) has shown a strong relation between judgments of the situation and decision making, particularly among experienced decision makers. In fact, the relation has been so strong that Klein has referred to his model of decision making as a "Recognition-Primed Decision (RPD)" model. Methodologically, Klein developed an extension of Flanagan's (1954) critical incident technique called the critical decision method (Klein, Calderwood, & MacGregor, 1989).

A pictorial representation of the RPD model is presented in Figure 2.2. It shows the proficient decision maker becoming aware of events that have occurred and relying on experience to recognize these events as largely typical. The judgments of typicality include guidance for plausible goals, critical cues, expectancies, and typical actions. The simplest case is one in which the situation is recognized and the obvious reaction is implemented. A more complex case is one in which the decision maker performs some conscious evaluation of the reaction, typically using imagery to uncover problems. The most complex case is one in which the evaluation reveals flaws requiring modifications, or the option is judged inadequate against one's goals and is rejected in favor of another option. The increased com-

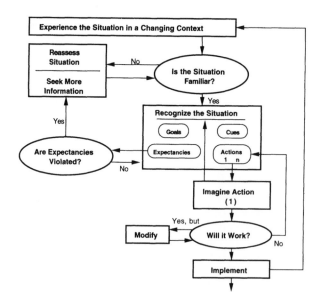

FIG. 2.2.   RPD Model.

plexity represented in these cases represents a movement from rule-based to knowledge-based processing (Rasmussen, 1986), a shift from predetermined actions to more conscious consideration of actions. In all cases, however, decision making is guided by the recognition (or understanding) of the situation. The failure to verify expectancies can alert the decision maker that the situation understanding is wrong, and that it is time to rethink it and to gather more information before preceding.

The strong coupling between judgment and decision, as represented, for example, by the Hypothesis and Option element's in Wohl's SHOR paradigm or the Situation and Action elements in Klein's RPD model, does not mean that decision making is necessarily easy. As Wohl (1981, p. 626) pointed out, there are "two realms of uncertainty in the decision-making process: (1) information input uncertainty, which creates the need for hypothesis generation and evaluation; and (2) consequence-of-action uncertainty, which creates the need for option generation and evaluation." It is this uncertainty, particularly under crisis or emergency-like contexts, that makes decision making "a worrisome thing, and one is liable to lose sleep over it" (Janis & Mann, 1977, p. 3).

Different cognitive processes become more or less important depending on where the uncertainty resides. For example,

Where options are more or less clearly prescribed but input data is of low quality (e.g., as in military intelligence analysis), a premium is placed upon creation and testing of hypotheses (e.g., where is the enemy and what is he doing?). Where input data are of high quality but options are open-ended (e.g., as in the Cuban missile crisis), a premium is placed upon creation and analysis of options and their potential consequences (e.g., if we bomb the missile sites or if we establish a full-fledged naval blockade, what will the Russians do?). . . . By contrast, tactical decision-making in support of combined air-land operations is generally characterized by both poor quality input data and open-ended options; hence, there is a much greater need than in other military situations for rapid hypothesis and option processing in the field. (Wohl, 1981, p. 626)

Uncertainty creates indecision and conflict. Hogarth (1987) presented a conflict model of decision making, a schematic of which is presented in Figure 2.3. Three notions are central to this model. First, people are goal directed. Second, "different forms of conflict are inherent in choice, whereby conflict is meant incompatibility" (p. 66). This conflict could be due to information-input uncertainty or consequence-of-action uncertainty, to use Wohl's terms. Third, conflicts are resolved by balancing the costs and benefits of alternative actions. This balancing could involve different alternatives, including whether to engage in or withdraw from the conflict itself. The "no-choice option" can occur anywhere in the decision process and leaves

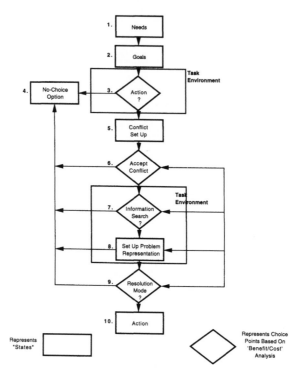

FIG. 2.3. Hogarth's Conflict Model. From Hogarth (1987). Reprinted with permission.

one with the status quo. According to Hogarth (p. 69), "this strategy is particularly likely to be followed under conditions of high uncertainty when unknown alternatives are difficult to evaluate."

At first glance, Hogarth's conflict model appears to be at odds with Klein's RPD model. In the former model, the decision maker seems to be in a constant state of inaction. In the latter model, the decision maker seems to be in a constant state of action, recognizing situations as familiar and immediately deciding what to do about them. Although the models do differ in their focus and emphasis (i.e., "conflict" vs. "recognition"), both appearances are too extreme.

Klein's RPD model has different levels of "recognition" and, therefore, delineates different cases under which decision makers generate and evaluate individual or multiple options. The different levels are determined by the decision maker's experience and task characteristics, such as time pressure, situational uncertainty, and the degree to which one's decision has to be justified after the fact. Similarly, the degree of conflict and, in turn, decision-making deliberation in Hogarth's model is dependent on one's situation understanding, referred to as the *task environment.* To quote Hogarth (p. 67):

> The choice to accept or reject the potential conflict in the situation takes place in a specific environment. Consequently, the way the environment is perceived at this early stage in the overall choice process can affect whether or not the individual will even attempt to make a choice.... The importance of the manner in which the task is initially perceived by the individual is thus emphasized and is indicated in Figure [2.3] by the dotted box enclosing box 3.

In sum, there is a tight coupling between judgments about the situation and the decisions one makes. Different paradigms may emphasize different parts of the coupling, as is illustrated by Klein's RPD model and Hogarth's conflict model. But the coupling is still part of the larger conception of human problem solving and goal-directed behavior, whether it is that of executives or the man in the street (Beach & Mitchell, 1987) or, for that matter, the rat in the runway (Tolman & Brunswik, 1935).

This discussion is not meant to imply that decision makers do not make errors of judgment or bad decisions. Humans are limited information processors who "satisfice," not "optimize" (Simon, 1955). Sometimes we make good judgments and decisions; sometimes we do not. Our critical concern is (a) what cognitive processes do humans employ, and (b) under what task conditions will they be successful or not? With that focus in mind, we consider first the research on judgment processes and, then, the research on decision processes. We conclude the review with a brief section overviewing the research on memory and attention.

JUDGMENT: INFERENCES ABOUT THE WORLD

This section overviews three areas of research about how people make judgments about the world: research performed within the context of Social Judgment Theory; research on cognitive heuristics and biases; and research on expertise. Each area is considered in turn.

Social Judgment Theory (SJT)

SJT was formulated by Kenneth R. Hammond and his colleagues and students and extends the theory of probabilistic functionalism developed by Egon Brunswik, a perception psychologist, to the realm of judgment and decision behavior. For a detailed review, the interested reader is referred to the paper by Hammond et al. (1975) and the edited volume in Hammond's honor by Brehmer and Joyce (1988). We focus here on the theoretical framework, task characteristics, and important findings relevant to people making judgments about the world. The application of SJT to decision making and interpersonal learning and agreement is considered in the section describing research on decision processes.

SJT represents the extension of Brunswik's lens model from perception to judgment, as presented in Figure 2.4. The goal of human judgment (Ys) is to accurately predict environmental criteria (Ye). These criteria (Ye) may be the causes of observed effects or future states of the world.

People use information or cues (Xi) to make judgments. The cues may be acquired from the immediate environment or from our memory of past events. The focus of our attention is critical when cues are acquired from the environment, for if one's attention is poorly focused, one cannot obtain the necessary information. Once acquired, the cues are processed (i.e., utilized) by the person according to heuristics or schemas (or scripts or other knowledge structures) to produce the judgment (Ys). As is discussed throughout this review, people utilize a large repertoire of heuristics and schemas to combine information into judgments. Moreover, in an effort to be accurate, people try to utilize the cue information ($r_{s,i}$) in a manner that matches the information's environmental validity ($r_{e,i}$). That is, people try to rely on valid predictors of environmental criteria (Ye).

SJT research has shown that specific task characteristics significantly affect people's ability to make accurate judgments. Before considering these findings, we first turn to discuss (a) the characteristics of judgment tasks studied within the SJT paradigm, and (b) the lens model equation, which has been developed to quantify the relation between the environmental and cognitive systems represented in Figure 2.4.

Task Characteristics. SJT research has focused on why judgment tasks can be difficult, even for experts. In particular, we focus on five task characteristics here. First, by themselves, individual cues are seldom perfectly valid predictors of Ye. Rather, the relationship between individual cues and the environmental criterion is typically probabilistic (i.e., $r_{e,i} < 1$). Consequently, any given piece of information is not, by itself, perfectly diagnostic of the event that caused it (or that which is to be predicted). This is compounded by the fact that, second, information is not always reliable. That is, from a test–retest perspective, the same conditions may not always result in the same cue values (e.g., X1 or X2 in Figure 2.4). This is often the case when the cue values are themselves judgments. It is important to note that reliability and validity are two different concepts. A piece of information can be 100% reliable and either totally diagnostic (100% validity) or undiagnostic (0% validity) in predicting Ye. However, the less reliable the information, the less valid it is because of the inherent uncertainty (i.e., error) in the information itself.

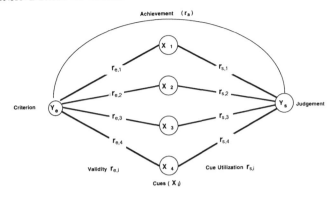

FIG. 2.4. The Lens Model.

In addition to the validity and reliability of the cues, the functional relations between individual cues and the criterion Ye may assume a variety of forms. For example, there may be a positive or negative linear relation between values on a cue and the criterion. Or, in contrast, there may be nonlinear relations, such as U-shaped, inverted U-shaped, or S functions. Substantial research has shown that nonlinear relations are difficult to learn and use in a consistent fashion.

Fourth, there are various organizing principles for combining the multiple pieces of information (cues) into a judgment. The principles that have been studied most are additive, averaging, and configural principles. Additive principles are those in which the independent contributions of the cues are totaled to make a judgment. For averaging principles, the contribution of each cue depends on the total set of cues present when making the judgment. By *configural principles* we mean that the person is looking for a pattern of cue values as indicative of one judgment or another. Pattern-matching principles have often been considered synonymous with schemas, which are knowledge representations that indicate how information should look (or go together) for different types of objects. But, as we discuss later, schemas are more than just expected data patterns.

Finally, overall task predictability when using all the cues is seldom perfect. In fact, many judgment tasks, such as predicting the stock market or success in school or the actions of other people, have low levels of predictability, even when one has all the information one wants. Poor predictive achievement in such tasks is not the fault of the person. He or she may be making all the sense one can from the data. The problem is in the task; there is simply not much predictive sense to be made. In sum, human judgment is not only difficult because of limited human information-processing capabilities, it is difficult because of the very nature of judgment tasks themselves.

The Lens Model Equation (LME) presented in equation [1] quantifies the relations between the environmental and cognitive systems represented in Figure 2.4:

$$r_a = GReRs + [(C1 - Re^2)(1 - Rs^2)]^{1/2} \qquad [1]$$

Its parameters are best described by beginning with an example. Specifically, the typical judgment task consists of presenting a person with a number of cases representing the judgment problem. Each case (or profile) for consideration consists of a mix of values on the several cues (Xi) being used to make a judgment (Ys) of the environmental criterion (Ye) for that case. For example, if one was making judgments (Ys) about the potential success of various stocks, one would do so by considering various pieces of information (Xi) about each stock, such as its price, price-to-earnings ratio, volatility, and so on. If one made good judgments, then there would be a high correlation between one's judgments (Ys) and the actual success of the stocks (Ye). Presumably, such an achievement would be due to (a) one's knowledge of the factors that predict successful stocks, and (b) the consistent application of that knowledge when picking stocks. However, it is well known that even the best stockbroker makes mistakes. These mistakes are due to the inherent unpredictability of the stock market itself.

This example illustrated all the parameters of the LME. We now consider the four parameter for equation [1] in turn: r_a is the correlation between the person's judgments (Ys) and the environmental criterion (Ye) over a number of cases. Statistically, it represents the person's accuracy in predicting the criterion Ye. In our example, it represented the stockbroker's accuracy in predicting stocks.

In most SJT research, G represents the correlation between the predictions of the best-fitting (using a least-squares procedure) linear model for predicting the person's judgments and the predictions for the best-fitting linear model for predicting the environment criterion. Conceptually, G represents the level of correspondence (or match) between the task and the person and is considered a measure of knowledge. To the extent that a person can accurately assess the diagnosticity of individual cues, their (linear) functional relations with Ye, and the organizing principles for combining cue information, the greater their knowledge (G) of the task and the better their performance (r_a). In our example, it was the stockbroker's knowledge of the factors the best predicted the success of stocks, as modeled by linear equations.

Rs is the correlation between the person's judgments and the predicted values of the judgments based on the (linear) model of the person. It specifically measures how well the linear model predicts the person's judgments. In those cases in which all the predictability can be captured by linear equations, Rs is typically considered to be the overall consistency or "cognitive control" exhibited by the person in making his or her judgment. It is analogous to assessing whether a person will give the same judgment for the same information seen a second time, although it is not a test–retest correlation. In our example, it was the consistency with which the stock-broker applied her or his knowledge about stocks.

Finally, Re is the correlation between the criterion and the predicted values of the criterion based on the (linear) environmental model. Consequently, it represents the overall linear predictability of the environmental system. In our example, it was the inherent predictability of stock success.

Finally, C is the correlation between the residuals for the best-fitting linear models for the environmental and person sides of the lens. A correlation that is significantly larger than 0.0 indicates that (a) there exists configurality in the data, such as patterns of cue values, (b) the person knows them, and (c) it increases overall achievement. Therefore, C also represents knowledge. From a modeling perspective, however, C represents knowledge of the configurality (i.e., the nonlinearity and nonadditivity) in the task, whereas G represents knowledge of the linearity in the task. Studies (e.g., see Johnson, 1988) have found significant values for the C correlation and thereby support for the argument that people are quite capable of learning and using non-linear data patterns when making judgments. Most of the variance in their judgments can be predicted, however, by using a linear, multiple regression equation, that is, the first term in the LME. As discussed later, these high levels of predictability are, in part, the result of the statistical robustness of the multiple regression equation. For this reason, as well as the tractability provided by the multiple regression equation for modeling the tradeoff process inherent in many types of judgments, only the linear component of the LME is used in many SJT studies.

Figure 2.5, from Balzer, Doherty, and O'Connor (1989), presents a version of the lens model that incorporates the components of the LME; r_a, G, and C are found at the bottom of the figure. Again, r_a is the correlation between Ye and Ys; G is the correlation between the predictions of the two linear, multiple regression equations; and C is the correlation between the residuals for these two equations. Re represents linear task predictability in terms of the correlation between Ye and predicted Ye. Rs represents linear predictability in terms of the correlation between Ys and predicted Ys.

The parameters of the LME provide important insights into the nature of many types of judgment tasks and people's ability to learn and perform them well, both within and outside of the laboratory. Before considering these insights, we make three points. First, G and C (i.e., knowledge) and Rs (i.e., cognitive control) are statistically independent. Consequently, it is possible to perform quite poorly, even if one knows what to do if one cannot do it consistently. Second, Re sets an upper bound on achievement, even if knowledge and cognitive control are perfect. In other words, task predictability is as important a determinant of judgment performance as our information-processing capability and/or the information systems that we develop to improve it. Third, judgment and decision making are tightly

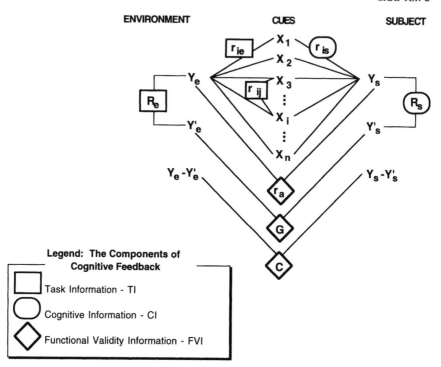

FIG. 2.5. The Lens Model incorporating the Lens Model equation (from Balzer et al., 1989).

coupled, but distinctly different concepts. For example, I might be a good judge of successful stocks, but decide not to invest, even when a good stock is available because of a cash-flow problem or a host of other reasons.

Important Findings. The major finding of overriding importance here is that task characteristics significantly affect people's ability to perform complex judgment tasks (r_a). Moreover, the LME permits one to assess whether the reason is poor knowledge (G) and/or cognitive control (Rs). First, we consider characteristics between individual cues and the criterion, such as the cue's ecological validity ($r_{e,i}$) or functional relations. Second, we consider the effects of task predictability (Re) on judgment.

Effect of Ecological Validities and Functional Relations. Early research within the SJT paradigm by Peterson, Hammond, and Summers (1965) showed that people had great difficulty in tracking changes in cue validities. Subsequent research by Summers (1969) showed that it was even more difficult for people to track changes in functional relations between cues and the criteria. Such changes are certainly possible, although one hopes infrequent, in the dynamic flow of events outside the laboratory.

At a more basic level, Hammond (1971) demonstrated that it was extremely difficult for people to learn nonlinear relations (e.g., inverted U-shaped functions) versus linear ones. Inverted U-shaped functions readily exist outside the laboratory. A nice example is provided by the story of Goldilocks and the Three Bears. Being too hot or too cold is not acceptable; it has to be just right. Moreover, Hammond and Summers (1972) showed that even after people learned the nonlinear and linear relations (i.e., G was the same), achievement (ra) was still lower for the nonlinear ones because of significantly lower cognitive control (Rs). That is, it was just harder for people to do what they wanted to do cognitively. In an effort to explain these findings, Brehmer, Kuylenstierna, and Liljergren (1974) presented data suggesting that people have a hierarchy of relations from which they sample when performing judgment tasks, and that the hierarchy goes from linear to nonlinear relations.

Research has shown that people have poor insight into their judgment process. For example, it has been consistently shown that people think they use more information than they actually do or that is required to accurately predict their judgments. In particular, people tend to underestimate the weight they place on important cues and overestimate the weight they place on unimportant cues when compared to the (highly predictive) weights obtained from a multiple regression analysis of their judgments (e.g., see Cook & Stewart, 1975). Moreover, experts appear just as susceptible to poor insight as novices (e.g., see Slovic, Fleissner, & Bauman, 1972). Anderson (1982) suggested that experts' lack of self-insight occurs because experts' judgment processes become more automatic with experience and, thus, less susceptible to conscious examination.

Part of the reason that various ecological validities and functional relations affect judgment is the form of feedback one typically receives in judgment tasks. This feedback is outcome feedback, that is, one makes a judgment and at some later time—often too much later—one finds out the outcome of it (i.e., Ye). SJT research has shown that outcome feedback is a poor form of feedback. One problem with it (which is discussed later) is that it is affected by task predictability. The more unpredictable the judgment task, the harder it is to both learn the judgment task (i.e., G) and to apply one's knowledge consistently (i.e., Rs). Second, outcome feedback is often delayed in time. Brehmer (1986) and Sterman (1989) showed that performance is significantly affected by such time delays. Lichtenstein, Fischhoff, and Phillips (1982) suggested that one of the reasons that meteorologists' judgments are so well calibrated to the probability of weather events is that they receive constant and fast feedback. And, third, in many tasks, such as selection tasks, one never even receives feedback about judgments and actions not taken (Einhorn & Hogarth, 1978).

Research has shown that cognitive feedback can significantly improve task performance. Cognitive feedback is oriented to providing people with information regarding how they use the cues to reach a judgment (e.g., their weights and function forms) and how they should use the information (e.g., the task weights and functions). Hammond et al. (1975) argued that one of the unique contributions of SJT research is the empirical demonstration that cognitive feedback results in significantly higher levels of performance than outcome feedback (i.e., receiving the correct answer).

The reason that cognitive feedback improves achievement is threefold. First, as discussed earlier, people have poor insight into how they actually use information to make a judgment. Second, they have poor insight into how they should use the information. In an extensive review of the literature, Balzer et al. (1989) showed that information about the task (i.e., the environmental side of the lens) is significantly more beneficial than information about the person (i.e., the cognitive side of the lens) for performance in complex judgment tasks. And third, as discussed later, people often get distracted by the lack of predictability (or noise) in the system. Cognitive feedback keeps them focused on important cue-criterion relations.

SJT research has extended the focus on cognitive feedback to group decision making. (This extension is considered in the section on decision making.) What is important to note here is that research has demonstrated that cognitive feedback of people's inference processes has significantly improved interpersonal understanding and subsequent performance over more traditional approaches, such as talking about the problem without cognitive feedback. Moreover, this has been extensively demonstrated in the laboratory (e.g., Hammond & Brehmer, 1973) and in real-world applications (e.g., see Hammond & Adelman, 1976; Hammond & Grassia, 1985; Rohrbaugh, 1988).

Task Predictability. We now turn to consider two general findings regarding the effect of task predictability. First, Adelman (1981) and numerous others have shown that cognitive control (Rs) and achievement (ra) are significantly decreased by decreasing task predictability (i.e., lowering Re). Yet, there is minimal (if any) affect on knowledge (G). This appears to occur for two reasons. First, people are typically able to learn the correct organizing principle, even under conditions of moderately low-task predictability. However, second, people act as if judgment tasks are perfectly predictable, that is, as if they have no error. Consequently, the lower the task's predictability, the more people try out new principles in an effort to predict Ye perfectly. Because they cannot do so because of the task unpredictability, they routinely return to the correct organizing principle. The result is the same level of knowledge over a large block of trials, but more inconsistency in judgment, that is, high G and low Rs in terms of the LME.

Second, as we tried to illustrate in our example of the stockbroker earlier, the overall predictability of the task can limit the overall level of performance, even for experts. Dawes and Corrigan (1974) illustrated this point using data from two principal domains: clinicians' predictions of neurosis versus psychosis using the MMPI test, and faculty members' predictions of students' success in graduate school based on students' grades, test scores, and other information. The average r_a achievement (correlation) in these domains was between .19 and .28. The task predictability of the optimal linear model (Re) varied from .46 to .69. As illustrated by the LME, it provides an upper limit on achievement given available data. Regardless of the experts' knowledge and whether or not they consistently applied their knowledge when evaluating the cases, the level of task predictability limited their overall level of achievement substantially.

Similar results have been obtained in other domains. For example, Speroff, Connors, and Dawson (1989) found an average achievement level of 0.42 for physicians' judgments of hemodynamic status in critically ill patients. Task predictability was .67, again putting a substantial limit on achievement. Ebert and Kruse (1978) found that security analysts' achievement levels varied between −.02 and .54 for various subsets of stocks. Task predictability ranged between .73 and .90. In this case there appears to have been substantial room for improving the experts' judgments.

A number of researchers have tried various ways to improve judgments. As the LME suggests, one way is to remove the inconsistency in the person's judgments, that is, make Rs = 1.0. That can be accomplished by using a model of how the expert makes judgments, instead of the expert, to make the judgments (Ys). This has been called *bootstrapping* by Goldberg (1970). He and others (e.g., Dawes & Corrigan, 1974; Ebert & Kruse, 1978) have shown that this significantly improves judgment, but not as much as possible.

A second way to improve judgment is to combine the judgments of the experts. As Hogarth (1978) pointed out, combining the judgments of experts increases accuracy because the experts' inconsistencies tend to cancel out each other. However, the composite achievement levels still tend to be below the optimal level possible, given the available data.

The third way to improve judgment is to develop the best-fitting model, given the data (Xi) and the criterion (Ye). That is, one develops the model on the environmental side of the lens that provides the highest levels of task predictability. Simple linear models typically perform as well as more complex nonlinear models for many judgment tasks. Such an approach depends on having actual criterion data (Ye) for a reasonably large number of cases. This is not often possible. Dawes (1979), Dawes and Corrigan (1974), and Einhorn and Hogarth (1975) showed that, for tasks with moderate to low levels of predictability and low correlations between the cues, the achievement level of models that (a) place equal weights on all the cues, and (b) have the correct sign for the linear functions relating the cues to the judgments, will often approach the achievement levels of optimal models and outperform most experts and their models. The reasons for this surprising finding have to do with the statistical robustness properties of linear models.

More recently, Blattberg and Hoch (1990) examined ways to combine the strengths of the model with the strengths of the expert. Although they identified many strengths (and weaknesses) of each, we focus on only two here. Specifically, the principal strength of the model is that it optimally integrates data that can be captured repeatedly. In contrast, the principal strength of experts is that they can take advantage of information that happens infrequently, but validly predict the criterion (Ye) when it occurs. This type of data was not available to the experts in the cases reviewed earlier. Blattberg and Hoch found that in cases in which both types of data were available for managerial forecasting problems, the experts' achievement levels were comparable to, and sometimes better than, that of the model. The highest achievement levels were obtained, however, by using the judgments of both the model and the experts. Even so, these achievement levels ranged from 0.50 to 0.90. Achievement was simply limited by the predictability of the task itself.

Summary. In SJT research, the quantitative and qualitative characteristics of the task, that is, the environmental side of the lens model, is used as a standard by which to compare human information processing, that is, the cognitive side of the lens. High levels of achievement depend on (a) knowledge (G), that is, matching our use of the information to how we should use it; (b) cognitive control (Rs), that is, implementing our knowledge in a consistent fashion; and (c) the level of task predictability (Re) given the available data.

A substantial amount of SJT research has focused on agreement among experts, as well as on achievement. This research is discussed in more detail in the section examining differences in judgment for different levels of expertise. In closing this section, we note that the types of tasks used in SJT research represent just some of the many different types of tasks that have been used by cognitive scientists studying human judgment. We now turn to research on cognitive heuristics.

Cognitive Heuristics

Cognitive heuristic research has used probability and statistical theory as a normative standard for evaluating human judgment performance. Probability and statistical theory represent powerful rules for logically combining uncertain information into a judgment. Moreover, many advanced information and decision systems use these rules. Therefore, from a cognitive engineering perspective, it is important to assess whether humans try to emulate these rules or use other types of processing heuristics. It is important to emphasize that we are not saying that people should use the rules of probability and statistical theory but, rather, whether they do so.

An overwhelming finding is that human decision behavior systematically deviates from (or is "biased" when compared to) a normative model that is assumed to be the optimal way to make the decision under investigation. To quote Hogarth (1987, p. 5), "People do not possess intuitive 'calculators' that allow them to make what one might call 'optimal' calculations. Rather, they use fairly simple procedures, rules or 'tricks' (sometimes called 'heuristics') in order to reduce mental effort."

Judgmental heuristics, and the biases they often spawn, are the results of human effort to understand and master the environment given limited information acquisition, retention, and processing capacities. For example, research has demonstrated that the perception of information is not comprehensive but selective. As Hogarth (1987) pointed out, it has been estimated that only about 1/70th of what is present in the visual field can be perceived at one time. How do we know what to select? The answer is that we anticipate information on the basis of our causal model of the environment.

Similarly, memory is limited; human beings recall but a small part of the information they initially acquire. Moreover, current theories support the view that memory does not access information in its original form but, rather, works by a process of associations that reconstruct past events and fragments of information that are typically consistent with one's causal model of the environment. Finally, research indicates that considerable mental activity involves processing information that is both acquired from the environment and recalled from memory to make judgments of probable cause. There are, however, significant differences in causal versus statistical (or normative) reasoning.

Kahneman, Slovic, and Tversky (1982) compiled an anthology of research studies demonstrating that, when compared to the tenets of probability and statistical theory, humans have limited appreciation for the concepts of randomness, statistical independence, sampling variability, data reliability, regression effects, and so on, when making probabilistic judgments. All of these concepts are important when making causal inferences. Although these concepts and cognitive heuristics can be represented in the lens model paradigm of SJT, the tasks used to investigate them and, in turn, orientation have been quite different in cognitive heuristics research. We first overview the kinds of tasks used in cognitive heuristics research and then consider important research findings.

Task Characteristics. The typical tasks used in cognitive heuristics research are simple word problems employing quantitative information that is optimally combined using the rules of probability and statistical theory. Von Winterfeldt (1988, p. 467) described the paradigm as follows:

1. A formal rule, known to the experimenter but not known or available to the subject is applied in formulating an intellectual task.
2. Subjects are asked to solve the task intuitively (without tools).
3. A systematic discrepancy between the formal rule and subjects' answers is found.

The following example is a classic problem studied by Tversky and Kahneman (1980), which is discussed in detail in Hogarth (1987, pp. 42–44):

A cab was involved in a hit-and-run accident at night. Two cab companies, the Green and the Blue, operate in the city. You are given the following data:

1. 85% of the cabs in the city are Green and 15% are Blue.
2. ,A witness identified the cab as a Blue cab. The court tested his ability to identify cabs under the appropriate visibility conditions. When presented with a sample of cabs (half of which were Blue and half of Green), the witness made correct identifications in 80% of the cases and erred in 20% of the cases.

Question: What is the probability that the cab involved in the accident was Blue rather than Green?

The typical response to this question is about 80%. This is consistent with the evidence that the witness correctly identified 80% of the cases in the court's test of his ability to correctly identify cabs. However, the correct answer to the problem is that the chance of the cab being Blue is 41%. The reason so many people are so far from the correct probability estimate, and even the correct color of the cab, is that they often fail to consider the a priori or base-rate information that 85% of the cabs in the city are Green, not Blue. Probability theory, and in particular Bayes' Theorem, requires that one systematically use both the base-rate information and the new evidence when making a probabilistic inference. Specifically,

$$p(\text{Blue} \mid \text{Say Blue}) = \frac{p(\text{Blue})\, p(\text{Say Blue} \mid \text{Blue})}{p(\text{Say Blue})} = \frac{(.15)(.8)}{.29} = .41$$

where p(Say Blue) = p(Blue) p(Say Blue | Blue) + p(Green) p(Say Blue | Green)
$$= (.15 \times .80) + (.85 \times .20) = .29$$

It is important to note that when the same base-rate information is given a causal meaning by rephrasing the problem to say "85% of accidents in the city involve Green cabs and 15% involve Blue cabs" the average response is 55%, much closer to that which is optimal according to Bayes' Theorem.

Important Findings. The preceding example illustrates that people give meaning to information in order to make causal sense out of it. Einhorn and Hogarth (1986) and Hogarth (1987) identified three critical aspects of causal reasoning. First, people make judgments of probable cause on the basis of a *causal field* that is basically analogous to a perceptual field. Judgmental like perceptual processes are attuned to differences; therefore, the relevance of potential causes depends on whether they are considered as differences in the problem context. Second, people use various imperfect indicators of causal relations called *cues-to-causality*. These cues include, for example, temporal order, covariation, contiguity in time and space, and similarity of cause and effect. And, third, the confidence people place in a causal explanation is affected by the extent to which they can imagine plausible scenarios for both it and alternative explanations. As Hogarth (1987, p. 161) pointed out in his discussion of creativity in problem solving, "use of both causal fields and the cues-to-causality help the mind establish order out of the mass of information with which it is confronted. On the other hand, this order is bought at the cost of being able to perceive alternative problem formulations (i.e., causal fields) and potential causal candidates."

Probability theory does not require causal reasoning, for probability theory is merely a set of rules that permit one to infer the relationship between probabilistic events if certain assumptions are met. In fact, Einhorn and Hogarth (1986) and Hogarth (1987) argued that the nature of causal reasoning not only differs in important respects from the dictates of probability theory, but that certain aspects of probability theory are antithetical to causal reasoning. As they pointed out, for example, causal reasoning is generally unidirectional (e.g., X causes Y). On the other hand, in statistical logic the relation between two events can be, and often is, discussed in either or both directions. For example, in order to use Bayes' Theorem to calculate the posterior probability of a hypothesis given new data or information (i.e., $P(H \mid D)$), one needs to assess the likelihood of the data given the hypothesis (i.e., $P(D \mid H)$). And, whereas statistical theory is based on the logical structure of information, causal reasoning is responsive to both structure and content in terms of the causal field, cues-to-causality, and the plausibility of alternative scenarios and causal explanations. In sum, although it may not be normatively correct when compared to probability and statistical theory, people use heuristics that weigh information on the basis of its perceived causal meaning, not its statistical diagnosticity.

The heuristics that humans use to attach meaning to information makes us susceptible to "biases," when compared to some normative (or presumed "optimal") standard, depending on how the problem is presented or "framed" (Tversky & Kahneman, 1981). This point was illustrated with the taxicab problem presented earlier. Bazerman (1990) argued that three general heuristics can account for many of the biases shown in the literature. The three heuristics are (a) the availability heuristic, (b) the representativeness heuristic, and (c) the anchoring and adjustment heuristic. Each is defined next using Bazerman's definitions (1990, pp. 6–7):

> *The Availability Heuristic.* [People] assess the frequency, probability, or likely causes of an event by the degree to which instances or occurrences of the event are readily "available" in memory. . . . [For example], the product manager bases her assessment of the probability of a new product's success on her recollection of the successes and failures of similar products during the recent past.

> *The Representativeness Heuristic.* [People] assess the likelihood of an event's occurrence by the similarity of that occurrence to the stereotypes of similar occurrences. . . . [For example, some managers] predict a person's performance based on the category of persons that the focal individual represents for them from their pasts.

> *Anchoring and Adjustment.* [People] make assessments by starting from an initial value and adjusting to yield a final decision. The initial value, or starting point, may be suggested from historical precedent, from the way the problem is presented, or from random information. For example, managers make salary decisions by adjusting from an employee's past year's salary.

Substantial psychological research has been performed trying to understand how cognitive heuristics result in biased judgment. Unfortunately, too much of this research has focused on the biases and not on the heuristics, or task conditions, that give rise to them. Indeed, the research on cognitive biases has often been presented as a cataloguing of human fallibilities. However, the cognitive heuristics that spawn these biases have both strengths and weaknesses.

On the positive side, heuristics permit humans with limited information acquisition, retention, and processing capabilities to not only establish order and meaning out of the mass of information with which they are confronted, but to develop new and creative ways of improving (and hopefully, never destroying) the environment. On the negative side, they expose limitations in reasoning when compared to normative models of judgment and decision behavior. This research is particularly important to the design of advanced information systems because it suggests that how information is presented by these systems can significantly affect the adequacy of our thinking. Moreover, probability and statistical theory are not esoteric concepts; they provide powerful rules for enhancing causal inference under conditions of uncertainty. These rules can be and have been used to improve unaided human decision behavior (e.g., see Kelly, Andriole, & Daly, 1981). The goal of cognitive engineering is the design of support systems that effectively combine human

and computer strengths and, thereby, improve human judgment in an interactive way.

In an effort to synthesize the research on cognitive heuristics and biases,
Hogarth (1987) catalogued cognitive biases according to the following four
information-processing stages: information acquisition, processing, output,
and feedback. The following subsections use this scheme. In addition, we
identify the heuristic or heuristics, for it is often the misapplication of two
heuristics that causes the bias. A list of the biases mentioned herein is
presented in Figure 2.6.

Biases in Information Acquisition. Hogarth (1987, p. 209) pointed out
that, "The issue of bias in information acquisition can be conceptualized by
enquiring when and why information becomes salient to an individual. This
question can be further broken down by noting that information can be
accessed from two sources: (1) the individual's memory; and (2) the task
environment." First, we consider cognitive biases that are memory based;
second, we consider those that are facilitated by task characteristics; and third,

- In "Information Acquisition"
    - Availability of Instances in Memory
    - Selective Perception of Information
    - Focus on Confirming (vs. Disconfirming) Information
    - Vividness (vs. Abstractness) of Information
    - Data Presentation
        - **Order Effects (e.g., Information Presented First or Last)**
        - **Intact Displays (vs. Sequential Presentation of Information)**
        - **Seemingly Logical Displays**

- In "Information Processing"
    - Conservatism Bias
    - Representativeness Heuristic
    - Conjunction Fallacy
    - Inconsistency
    - Law of Small Numbers
    - Regression Bias
    - Task Characteristics Affecting Processes
        - **Time Pressure**
        - **Social Pressure**
        - **Information Overload**

- In "Output"
    - Question Format
    - Scale Effects

- In "Feedback"
    - Task Characteristics ("Outcome Irrelevant Learning Structures")
    - Misperceptions of Chance
    - Logical Fallacies in Recall
    - Hindsight Bias

FIG. 2.6. Cognitive heuristics and biases. From Hogarth (1987). Reprinted
with permission.

we consider biases that are facilitated by the way that data are presented, which is a particularly relevant task characteristic for displays and interaction routines.

We consider three memory-based biases that affect data acquisition. The first is the availability bias (also called the "ease of recall" bias by Bazerman, 1990, p. 40) resulting from the application of the availability heuristic (e.g., see Tversky & Kahneman, 1973). As was noted earlier, the ease with which specific instances can be recalled from memory affects judgments of the frequency of different events. Publicity or extensive discussion of (or focus on) particular events makes them more salient and, in turn, more available in memory. Availability does not, however, make them more accurate. Numerous irrelevant factors, such as the vividness of the event or the effects of one's imagination, can make an event more memorable and thus more available, without having any effect on its accuracy.

The second memory-based bias affecting information acquisition is selective perception (e.g., see Dearborn & Simon, 1958). In particular, people structure problems on the basis of their own experience such that anticipations of what one expects sometimes bias what one does see. Availability and representativeness heuristics appear to be operating here. One's experiences are more available in memory and thus more readily brought to bear on the problem. But they are not brought to bear in a haphazard way. Rather, they are brought to bear in a manner representative of one's world view, that is, according to one's cognitive model (to use SJT terms) of how cues go together in the environment.

One's cognitive model represents a reference point; we expect to see a pattern of data consistent with it. This is not bad. Quite the contrary, for accurate cognitive models permit us to go efficiently beyond the current data to make accurate judgments. Sometimes, however, our model and expectations are too strong. They cause us to perceive only what we expect to see or can make sense of, and this results in biased judgments.

The third memory-based bias, which is related to selective perception, is the confirmation bias (e.g., see Wason, 1960). In particular, people seek information consistent with their own views and hypotheses instead of seeking disconfirming information. Confirmation bias has been observed for a wide range of tasks, including those performed by experts, for example, Army tactical intelligence analysts (Tolcott, Marvin, & Lehner, 1989) and Navy sonar analysts (Cohen & Hull, 1990). This bias can be caused by an anchoring and adjustment heuristic in which one overweighs prior information, as suggested by Tolcott et al. However, it also can be caused by the misapplication of the representativeness heuristic, that is, one looks for the data that are consistent with (or representative of) one's cognitive model of the environment. Said differently, we try to prove our model right. This approach is at odds with the formal logical perspective in which we should try to disprove our model, that is, seek disconfirming evidence. Although Klayman and Ha (1987) have shown that the appropriateness of different test strategies depends on task characteristics, they concluded that people often are not sensitive to these characteristics and use a "positive test strategy as a general default heuristic" (p. 225).

We now consider two task-based biases affecting information acquisition. The first is the frequency bias. Specifically, people often judge the strength of predictive relations by focusing on the observed frequency of events rather than on their observed relative frequency. As Einhorn and Hogarth (1978) have shown, information on the nonoccurrence of an event is often unavailable and, more importantly, frequently ignored when available. Yet, such information is critical to making accurate probability estimates.

Second, concrete information, that is, information that is vivid or based on experience or incidents, dominates abstract information, such as summaries and statistical base rates. According to Nisbett and Ross (1980), concrete and vivid information contributes to the "imaginability" of the information and, in turn, enhances its impact on judgments. Yet highly vivid information may cause one to ignore other less vivid but essential parameters for accurate probability estimates, such as base rates. For example, a discussion with one's neighbors about the problems they have been having with their car may make one forget about the base rate of excellent performance for the car described in *Consumer Reports*. (Note: Failing to consider base rates also illustrates the application of the representativeness heuristic. This case is considered later in the subsection on information-processing biases.)

There is a substantial amount of research suggesting that the way information is acquired can significantly affect one's judgments. We give two examples. First, substantial research has shown that one can obtain different judgments for the same information by sequentially presenting the information in different orders. This has been referred to as an order-effects bias. Einhorn and Hogarth (1987) and Hogarth and Einhorn (1992) argued that this occurs because people use an anchoring and adjustment heuristic to process the information. In particular, they argued that people anchor on the current position and then adjust their belief on the basis of how strongly the new information confirms or disconfirms the current position. Moreover, the larger the anchor (i.e., the stronger one's belief in a hypothesis), the greater is the impact of the same piece of disconfirming information. Conversely, the smaller the anchor, the greater is the impact of the same piece of confirming information. According to the Hogarth–Einhorn model, this differential weighing of the most recent piece of information results in different anchors from which the adjustment process proceeds and, in turn, different final probabilities based on the sequential order of the same confirming and disconfirming information.

Adelman and Bresnick (1992), Adelman, Tolcott, and Bresnick (1993), and Ashton and Ashton (1990) found support for the anchoring and adjustment heuristic proposed by Hogarth and Einhorn to explain order effects. In addition, Adelman and Bresnick (1992) found support for a representativeness heuristic (via pattern-matching) in certain cases. This again suggests that people use multiple heuristics depending on task characteristics. It is important to note here that these studies were performed with experts. Moreover, Adelman and Bresnick (1992) found that information order not only affected Patriot air defense officers' judgments about the likelihood that an unknown aircraft was friendly or enemy, it also affected their decision to engage the aircraft or not. Chapter 7 presents a more recent study investigating order effects.

Fischhoff, Slovic, and Lichtenstein (1978) provided the second example of how the way information is acquired can affect judgment. Specifically, they demonstrated that seemingly complete presentations of information via logical displays can blind experts, in this case, auto mechanics, to critical omissions in the data. "All information presented seems so consistent that the individual is only able to come to one, possibly erroneous conclusion" (Hogarth, 1987, p. 211). Such an effect is consistent with the confirmation bias and the effect of both the availability and representativeness heuristics. More generally, the way information is presented affects its saliency and, in turn, the information that is acquired to make causal sense out of the world.

Biases in Information Processing. Related to information acquisition is its processing, that is, the heuristics, rules, or, more generally, ways that different pieces of information are combined into an inference or decision. Consistent with the paradigm for cognitive heuristics research, the literature is replete with processing biases compared with normative combination rules. Some of the biases are listed next.

Humans are conservative information processors compared to Bayes' Theorem (e.g., see Edwards, 1968). That is, even when the task is structured to make us focus on base rates, we do not revise our opinions on the receipt of new information as much as we should, compared to Bayes' Theorem. This is not surprising, for Bayes' Theorem is neither intuitively obvious nor a trivial analytical calculation. Instead, we appear to use two very simple heuristics to revise probability estimates on the basis of new data. First, there is the representativeness heuristic (e.g., see Kahneman & Tversky, 1973), which was illustrated in the cab example. That is, we sometimes judge the likelihood of events by estimating their similarity to the class of events of which they are supposed to be an exemplar, such as the witness's 80% accuracy in the cab example. The second heuristic is anchoring and adjustment. That is, predictions are made by anchoring on a value and then adjusting for the circumstances of the present case. Einhorn and Hogarth (1985) presented empirical results supporting the use of anchoring and adjustment when people make probability estimates in the face of highly ambiguous data, as well as in the case of different information orders considered earlier.

A number of other processing biases relevant to causal inference have been identified in the literature. For example, there is the conjunction fallacy (e.g., Slovic, Fischhoff, & Lichtenstein, 1976). In probability theory, the joint probability of two events cannot be larger than the probability of the smaller of two events (e.g., $.1 \times 1.0 = .1$). Yet, substantial research shows that when word problems are framed to elicit causal reasoning instead of statistical reasoning, people estimate the joint probability of two events as being larger, not smaller, than the probability of the smaller event. In one experiment, Tversky and Kahneman (1983) asked (in 1982, p. 123) professional forecasters to assess the probability of "a complete suspension of diplomatic relations between the USA and the Soviet Union sometime in 1983." A second group of forecasters was asked to evaluate this outcome and "a Russian invasion of Poland." The second group's probabilities were higher than the first's, thereby violating the laws of probability theory.

As discussed under SJT research, people are often inconsistent in how they process information. In addition, we are subject to what Tversky and Kahneman (1971) called the "law of small numbers" in contrast to the Law of Large Numbers. In particular, problems can be framed in such a way that people, including trained scientists, can give undue confidence to a (relatively) small amount of data. We are also subject to regression biases. To quote Hogarth (1987, p. 39):

> Extreme values of cues [without perfect predictability] (e.g., test scores) are typically accompanied by less extreme values of the criterion (e.g., performance measures). Thus, a sensible judgmental strategy is to regress predictions based on extreme observations toward the mean of the variable predicted. However, people frequently make implicit use of extreme values of predictive cues, together with consistency of data sources, to justify confidence in their predictions. Thus, paradoxically, characteristics of information that inspire confidence are often inversely related to the predictive accuracy of that information. This has led to what Kahneman and Tversky (1973) have termed the "illusion of validity."

In all these cases, people focus on the representativeness of the data instead of employing the more abstract concepts in probability theory.

In closing this subsection, it is important to note that task characteristics can significantly affect causal inference. In particular, Payne, Bettman, and Johnson (1987) showed that time pressure significantly affects the amount of information examined and the rules used to process it. Janis's (1972) classic work on groupthink shows how social pressure within highly cohesive, strongly directed groups can unduly influence judgment. And, as Janis and Mann (1977, p. 17) pointed out, "When the degree of complexity of an issue exceeds the limits of cognitive abilities, there is a marked decrease in adequacy of information processing as a direct effect of information overload and ensuing fatigue."

Biases Due to Output or Feedback. As Hogarth (1987, p. 213) pointed out, " 'Output' biases appear to be triggered by the way in which people express judgement or choice. . . . The importance of feedback in judgement relates to its effect on learning." Regarding the former, Hogarth (1975) showed, for example, that probability estimates can depend on how people have been asked to respond and on the scale used to measure these responses. This output bias is so strong that Spetzler and Stael von Holstein (1975) strongly recommended that decision analysts use multiple methods to converge on probabilistic assessments of uncertainty.

The reader should not think that this output bias is simply a function of using numbers instead of words to represent uncertainty. Moore (1977) found substantial differences in how experienced managers ranked expressions used to express uncertainty. And in the domain of intelligence analysis, Kent (1964) "was so concerned with the fact that almost all intelligence analysis documents contained ambiguous verbal reports of uncertainty, that he proposed a set of rules for translating words into probabilities and vice versa" (von Winterfeldt, 1988, p. 465).

More generally, as Fischhoff, Slovic, and Lichtenstein (1980) showed, the way a person is asked to respond can significantly affect his or her judgment. In addition, it can affect one's decision making. For example, the relative preference for gambles can be reversed when people are asked to express choices in different ways. Moreover, such preference reversals have even been demonstrated in a Las Vegas gambling casino (Lichtenstein & Slovic, 1973) and, more recently, with computer-based information displays (Johnson, Payne, & Bettman, 1988).

Feedback concerning the outcomes of our judgment can also induce bias. This might occur either as a result of (a) how we interpret outcomes, or (b) the nature of the environment itself. For example, regarding the former, people have a tendency to attribute success to their skill and failure to either chance or the situation with which they were faced. Ironically, people tend to attribute other people's failures to personality traits, not to the situation. Such success or failure attributions appear to be part of a more "fundamental attribution error" (e.g., see Nisbett & Ross, 1980, p. 31) in which people tend "to ignore powerful situational determinants of behavior." And, considering the latter, outcomes often yield inaccurate or incomplete information concerning predictive relations.

Regarding the effects of the environment, SJT research clearly indicates that outcome feedback does not necessarily result in high levels of performance. Part of the problem is that inference tasks are seldom perfectly predictable (i.e., Re < 1); consequently, outcome feedback is seldom of perfect validity. In addition, as Einhorn and Hogarth (1978) pointed out, in most decision-making settings, one can seldom (if ever) learn how good one's judgment is (e.g., as operationalized statistically by a correlation) because one's decisions and subsequent actions make it impossible to know what the effect of other decisions and actions would have been. In addition, feedback is often delayed or affected by other events. In particular, research on dynamic decision making (e.g., see Sterman, 1989) has shown that delays in receiving outcome feedback can result in misperceptions that essentially make people insensitive to it. Such settings have been referred to as "outcome irrelevant learning structures" by Hogarth (1987, p. 222).

In addition to task characteristics, people are not always effective in using feedback. For example, people often misperceive chance fluctuations and, in some cases, erroneously attribute causation to chance events. A good example is the "gambler's fallacy" in which people expect a chance event (e.g., Black in roulette) because they have just observed a large number of unexpected (but chance) events (e.g., a number of Reds). Again, this is an example of the use of the representativeness heuristic.

People often cannot remember all the feedback they receive. This can result in logical fallacies in recall. In studying eyewitness testimony, Loftus (1979) showed that recall can be influenced by postevent information and the way information is elicited from the witness. This appears to occur because memory is based on the reconstruction of events to make causal sense out of the world. To quote Hogarth (1987, p. 134), "the paradox of memory is that long-term memory does not work by remembering what is actually recalled, but rather by remembering fragments of information that allow one to construct more complete representations of the information." Seen from this perspective, the hindsight bias, which is colloquially referred to as Monday morning quarterbacking or second guessing, is not surprising. Perhaps what is surprising is its strength for, as Fischhoff (1975) found, people appear adept at interpreting new information as consistent with previously held positions.

Research on Expertise

Much of the research reviewed in the previous two subsections has shown significant limitations in expert judgment. First, the predictive accuracy (r_a) of experts in many domains is surprisingly low because of the low levels of task predictability (Re) in those domains. Multiple regression equations based on the criterion data (Ye) routinely outpredict the experts. Sometimes, even simple linear models with equal weights on the cues outpredict them. Although Blattberg and Hoch (1990) showed that expert judgment can be improved substantially in those cases in which experts have access to reliable cues that are unavailable to the model, "expert plus model" performance is often still surprisingly low.

Second, the heuristics and biases literature has shown that experts are susceptible to using incorrect heuristics with the result being biased judgments. For example, Adelman and Bresnick (1992) and Ashton and Ashton (1988, 1990) showed that Army air defense officers and auditors, respectively, are susceptible to order-effects biases due to the use of an anchoring and adjustment heuristic. Northcraft and Neale (1987) showed that realtors' property appraisals can be affected significantly by the listing price due to their use of an anchoring and adjustment heuristic. And Tversky and Kahneman (1971, 1983) showed that statisticians and forecasters, respectively, are susceptible to the "law of small numbers" and the conjunction fallacy if the problem is framed in a way that induced them to rely on the representativeness heuristic.

These findings are not meant to imply that knowledge is not an inherent aspect of "expertise" or that it is not essential for high performance levels in many disciplines, such as physics, chemistry, chess, and so on. Obviously, it is. In fact, considerable cognitive research has examined how experts acquire, organize, and use knowledge when solving problems in these disciplines. Moreover, it has examined how experts differ from novices in these regards. This is the research literature we review in this subsection, for it is critical when cognitively engineering technology for different users.

We begin by reviewing the results of cognitive research performed in disciplines such as physics, chemistry, and chess. Then, we attempt to reconcile the discrepant findings regarding experts' predictive accuracy. We note here that research by Camerer and Johnson (1991) and Johnson (1988) suggests that experts employ similar structures and processes in disciplines in which they perform well and poorly, and that it is task characteristics (e.g., Re) that determine performance. Although Hammond (1988; Hammond, Hamm, Grassia, & Pearson, 1987) and others (e.g., Holyoak, 1991) also have emphasized the critical importance of task characteristics for high performance levels, they have further argued that judgment processes are themselves task dependent. Such a position is the essence of a cognitive engineering perspective, for it implies a direct relationship between a system's design and the cognitive processing of its users.

Task Characteristics. Ericsson and Smith (1991) defined three steps as part of the "expertise" approach to understanding differences between expert and novice problem solvers in various domains. The first step is to "identify or design a collection of representative tasks to capture the relevant aspects of superior performance in a domain and to elicit superior performance under laboratory conditions" (p. 8). The key ideas here are control and representativeness. That is, one must have sufficient control in the

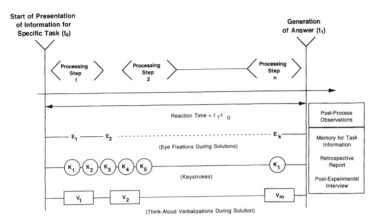

FIG. 2.7. Alternative data collection measures.

laboratory such that the tasks one uses will reliably elicit performance differences between experts and novices. In addition, the performance differences elicited in the laboratory must be representative of those observed in actual settings.

The second step involves the analysis of superior and inferior performance of experts and novices, respectively. This analysis is directed toward the third step, which is being able to account for performance differences. Much of our discussion focuses on the third step, that is, the "important findings" regarding the different knowledge structures and cognitive processing of experts and novices. This understanding is, of course, based on our methods for data collection and analysis in step 2. One cannot directly observe cognitive processes, but must infer them based on the data we collect for various measures.

Figure 2.7 (from Ericsson & Smith, 1991) shows different kinds of data collection measures for an illustrative study. Four measures were collected while the expert (or novice) was performing the task: reaction times, eye fixations, keystrokes, and think-aloud verbalizations. Three additional measures were obtained after the task was completed: memory for specific task information, retrospective reports in which participants stated how they were performing the task, and interviews after the experiment was over. Data for these measures then were used to infer how participants processed the data while performing the task.

All studies do not use the same measures. One reason is that different studies emphasize different aspects of expertise, for example, memory versus reasoning processes. A second reason is that some measures are typically given more credence than others. For example, think-aloud verbalizations are better than retrospective reports (Ericsson & Simon, 1984; Nisbett & Wilson, 1977). As Hammond (1976, p. 4) pointed out:

> Verbal representations of introspective reports of covert mental operations are poor representations because of the well-known inaccuracy of introspection; that is, despite the best of intentions, any introspection regarding the basis for a judgment may simply be an incorrect description of the cognitive activity that took place. . . . [Moreover], persons making the introspection will not have the conceptual or technical ability required to describe their judgment processes completely; indeed they will not even know what a complete description should consist of.

The typical analysis method using think-aloud verbalizations has been called *protocol analysis*. The audiotapes (or protocols) containing the verbalizations are typed up for subsequent analysis, typically by constructing problem behavior graphs (Newell & Simon, 1972). Problem behavior graphs resemble flowcharts composed of nodes (knowledge states) connected by arrows (process operations). Often, these graphs are compared to the results of a task analysis performed prior to the study. The task analysis involves specifying the different processing sequences that could generate the correct answer given the participants' preexisting knowledge. One then can compare the problem behavior graphs with the task analysis sequences to determine whether participants with different levels of expertise employ different processing sequences (Ericsson & Smith, 1991).

There are, of course, other methods for analyzing think-aloud verbalizations. For example, Adelman, Gualtieri, and Stanford (in press), Isenberg (1986), and Schweiger et al. (1985) coded verbalizations into categories representing different cognitive processes in an effort to understand which processes were employed most frequently during problem solving and how task characteristics affected them. Cohen and Hull (1990) employed a correlational approach to investigate the relation between verbalizations representing critical events in a sonar classification task.

Similarly, there are other methods for obtaining process behavior beside think-aloud verbalizations or the alternative methods listed in Figure 2.7. For example, Patel and Groen (1991) developed a method called *diagnostic explanation*. In this process, participants are asked to explain the causal patterns underlying a clinical case. From this explanation, Patel and Groen developed a semantic network to describe the causal and conditional rules in the participant's explanation. A semantic network is a graph formed by nodes and connecting paths and, therefore, is quite similar to a problem behavior graph. The innovation is not in the analysis procedure, but in the data collection procedure.

Another data collection procedure is the mouselab system developed by Johnson, Schkade, and Bettman (1989). The mouselab employs a keystroke method in which one uses a mouse to assess information on a computer screen. The information is arranged in a matrix format and concealed until the participant opens selected boxes with the mouse. The matrix has typically been an information item × decision alternative matrix, for the mouselab primarily has been used to study decision behavior. Kirschenbaum (in press), however, recently used an item × time matrix to examine the information acquisition and processing strategies of naval officers varying in expertise.

A third data collection method is the Critical Decision Method (CDM) developed by Klein and his associates (e.g., see Klein et al., 1989). The CDM is a retrospective interview strategy extending the Critical Incident Technique developed by Flanagan (1954). Like the earlier technique, the CDM begins with the person selecting an actual, nonroutine incident requiring expert judgment or decision making. Once the incident is selected, the interviewer asks for a brief description. Then, and unlike the earlier technique, a semistructured format is used to probe different aspects of the judgment and decision process. These aspects include, for example, the cues and knowledge used in the situation, previous experiences (i.e., analogues) that came to mind, and the goals, options, and reasoning involved in the judgment and decision process. It is through this cognitive probing that one tries to understand the knowledge structures and processing strategies that distinguish experts from novices.

The CDM is the one technique that routinely has been used outside the laboratory in naturalistic settings, such as fire fighting, Army command and control, and newborn care. It violates the first step in Ericsson and Smith's (1991) three-step approach to studying expertise because the participant, not the researcher, selects the task; consequently, there is no control over the material presented to the participants. Nevertheless, the data collected thus far suggest that the CDM is a reliable and valid method for use outside the laboratory. Klein et al. (1989) have obtained high levels of interrater agreement for coded information taken from transcripts over a wide range of domains. Moreover, they have obtained comparable results when using more traditional laboratory tasks. Taken together, this supports the viability of the CDM approach for understanding expertise.

Important Findings. We now turn to consider some of the important findings of the research on expertise. We begin with the listing of findings presented in Glaser and Chi (1988) and then move toward a consideration of knowledge structures and cognitive processes as a means of integrating the results.

1. *Experts Excel Mainly in Their Own Domain.* The obvious reason is that they have more domain knowledge than novices. In addition, they have both a larger repertoire and more powerful strategies for processing information in their own domain.

2. *Experts Perceive Large Meaningful Patterns in Their Own Domain.* "It should be pointed out, however, that this ability to see meaningful patterns

does not reflect a generally superior perceptual ability; rather it reflects an organization of the knowledge base" (Glaser & Chi, 1988, p. xvii). That is, experts know what information should go together; consequently, they can recognize and recall meaningful patterns when they see them, whereas the novice cannot. However, experts are no better than nonexperts at recognizing or recalling random patterns.

3. Experts are Faster than Novices at Performing the Skills of Their Domain, and They Quickly Solve Problems with Little Error. As Glaser and Chi emphasized, there are at least two ways to explain experts' speed. The first is that, due to many hours of practice, the skill is more automatic. Expert typists, for example, are faster than novices, in part, because they can move their fingers faster. Second, experts can arrive at solutions without conducting extensive search. "The patterns that chess experts see on the board suggest reasonable moves directly, presumably because, through many hours of playing, they have stored straightforward condition-action rules in which a specific pattern (the condition) will trigger a stereotypic sequence of moves" (p. xviii). However, as the research reviewed in the section on Social Judgment Theory indicates, task characteristics are a principal determinant of the accuracy of expert judgment.

4. Experts Have Superior Short-Term and Long-Term Memory. Chase and Ericsson (1981) proposed a model of "skilled memory" to account for these findings. The model asserts that improved short-term memory is the result of a more efficient use of long-term memory. In particular, experts use existing semantic knowledge and patterns to generate long-term memory encodings of the presented information. These encodings include the retrieval cues for recalling the information.

5. Experts See and Represent Problems in Their Domain at a Deeper (More Principled) Level Than Novices. As Glaser and Chi (1988, p. xix) pointed out, "both experts and novices have conceptual categories, but the experts' categories are semantically or principle-based, whereas the categories of the novices are syntactically or surface-feature oriented." For example, when solving physics problems, experts use the principles of mechanics to organize their thinking, whereas novices use the objects stated in the problem description.

6. Experts Spend Time Analyzing a Problem Qualitatively. "Protocols show that, at the beginning of a problem-solving episode, experts typically try to 'understand' a problem, whereas novices plunge immediately into attempting to apply equations and to solve for an unknown. What do the experts do when they solve for an unknown? Basically, they build a mental representation from which they can infer relations that can define the situation, and they add constraints to the problem. . . . Adding constraints, in effect, reduced the search space" (Glaser & Chi, 1988, p. xix).

7. *Experts Have Stronger Self-Monitoring Skills.* For problems such as physics and chess, studies have found that experts are more aware than novices of when they have made an error or when they need to check their solution or, more generally, when they have failed to comprehend something. This enhanced self-monitoring ability is a function of greater domain knowledge and their more principle-based mental representation of the domain. (Note: The research performed within the Social Judgment Theory and heuristics and biases paradigms raises questions about the generality of these findings to other types of tasks, particularly to those with low task predictability and outcome irrelevant learning structures.)

In all cases, experts' superior performance is accounted for by enhanced domain knowledge and mental representations of that knowledge. Many different constructs have been discussed for describing different knowledge representations or structures. These constructs include if–then (condition–action) rules, schemas, frames, scripts, images, and mental models. However, the differences between these constructs have been operationalized best in different types of expert system software, not in terms of cognitive processes. Regarding the latter, for example, Beach (1990) seemed to use the terms *schemata, frames, models,* and *images* almost interchangeably. Noble's (1989, p. 474) schema theory was derived from "the frame descriptions of Minsky (1975), the script theory of Schank and Abelson (1977), and the schema theory of Rumelhart (1975)." And many of the papers in recent workshops on mental models (e.g., Rogers, Rutherford, & Bibby, 1992), including a workshop focusing on "mental models and user-centered design" (Turner, 1990), attempted to develop a distinct operationalization for the mental model construct. Finally, it is important to reiterate that Social Judgment Theory uses the phrases "judgment model" and "model of the cognitive system" when referencing the person side of the lens model.

As Cohen (1991, p. 48) pointed out, there is a certain sense of arbitrariness in distinguishing between different knowledge structures:

> Confusion in this area, we think, is due to misunderstanding the role of knowledge structure. Its importance is not tied to specific hypotheses about human cognitive architecture; nor does it entail the choice of a particular knowledge representation language or expert system implementation. Structure is not a syntactic unit or format, e.g., frame, an assertion, a rule; any of these devices might be used, interchangeably, to represent structure. Rather, structure refers to the pattern of inferences that are possible in a domain. It refers to what Newell [1981] calls the 'knowledge level;' what the person knows, not the details of the mechanisms by means of which he or she knows it. In the important case of causal structure, it means locality of influence, i.e., a particular event or variable can be understood and predicted (to a reasonable approximation) by reference to a particular set of objects, events, and variables.

It is for these reasons that we have chosen to use the more general terms of *mental representations* and *mental models*. As Rogers et al. (1992) pointed out, Craik's (1943) original thesis of mental models was that they were "internal constructions of some aspect of the external world that can be manipulated enabling predictions and inferences to be made" (p. 2). From this perspective, experts have more knowledge and sophisticated models than novices about the external world pertaining to their domain. They know what information should go together and, therefore, are better able than novices to (a) recognize meaningful patterns in the data, (b) remember those patterns and relevant information when they see it, and (c) know what new information is most important to obtain or what action should be taken next, that is, how to best go beyond the current data. Similarly, their more sophisticated models let them, if need be, (d) represent the problem at a deeper, more principled level; (e) employ a range of processing strategies; and (f) better monitor their progress by mentally simulating the effectiveness of different solution paths.

The superiority of experts' mental representation of their domain, and their apply to efficiently perceive, remember, and process information using this representation, appears to be common across domains. We quote Johnson (1988, p. 217), who studied physicians evaluating medical students applying for admissions into hospital internships and residencies:

> In sum, the experts appear to examine this information in a top-down fashion, using their knowledge of medical education to structure their search. The increased use of goals in their protocols, along with the greater use of knowledge retrieved from memory and their more active search patterns, presents a picture consistent with the portrait of experts in other domains. Our experts appear to use their knowledge to examine only information that they consider diagnostic, limiting their search to a smaller subset of the available information. Thus, although admissions tasks such as this have often resulted in rather disappointing evaluations of expert performance, these results suggest that the processes used by these experts are much like those used by experts in other domains.

In a subsequent paper, Camerer and Johnson (1991, pp. 207–212) asked two questions of principal importance here. The first question was, "Why do experts develop configural rules?" A *configural rule* is defined as one in which the impact of one variable (or cue in Social Judgment Theory) depends on the value of other variables. The use of configural rules or *pattern matching* is quite common among experts because of the highly interrelated nature of their mental representations of the domain. For example, the if–then rules used in many expert systems are typically configural rules specifying the conditions under which one can take specific actions. The second question was, "Why are configural rules often inaccurate?" As emphasized previously, the appropriateness of this question depends on the predictability of task domain under consideration. In addition to making this point, Camerer and Johnson discussed the role of heuristics and biases.

First, configural rules are easy to learn and use. They are typically suggested by prior theory and are easy to weave into a causal narrative: "These coherent narratives cement a dependence between variables that is easy to express but may overweight these 'causal cues', at the cost of ignoring others. Linear combinations yield no such coherence" (p. 208). In addition, configural rules emerge naturally when trying to explain past cases. Most importantly:

> Inventing special cases is an important mechanism for learning in more complex deterministic environments, where it can be quite effective. The tendency of decision-makers to build special-case rules mirrors more adaptive processes of induction (e.g., Holland, Holyoak, Nisbett, & Thagard, 1986) that can lead to increased accuracy. As Holland and his associates pointed out, however, the validity of these mechanisms rests on the ability to check each specialization on many cases. In noisy domains like the ones we are discussing [that is, ones with low task predictability, Re], there are few replications. (Camerer & Johnson, 1991, pp. 208–209)

This naturally leads to the second question, why are configural rules often inaccurate? We again quote Camerer and Johnson (1991, pp. 209–210) at length:

> One reason configural rules may be inaccurate is that whereas they are induced under specific and often rare conditions, they may well be applied to a larger set of cases. Often, people induce such rules from observation, they will be overgeneralizing from a small sample (expecting the sample to be more 'representative' of a population that it is—Tversky & Kahneman, 1982). Configural rules may also be wrong because the implicit theories that underlie them are wrong. A large literature on 'illusory correlation' contains many examples of variables that are thought to be correlated with outcomes (because they are similar) but are not. Inaccurate configural rules may persist because experts who get slow, infrequent, or unclear feedback will not learn that their rules are wrong. When feedback must be sought, inaccurate rules may persist because people tend to search instinctively for evidence that will confirm prior theories (Klayman & Ha, 1987). Even when feedback is naturally provided, rather than sought, confirming evidence is more retrievable or 'available' than disconfirming evidence (Tversky & Kahneman, 1973, p. 201).

In their answer to the question of why inaccurate configural rules persist, Camerer and Johnson referred to at least three biases: representativeness, confirmation, and availability. More importantly, their answer implicitly focused on aspects of the task domain: noisy (as contrasted with deterministic); filled with rare conditions, inaccurate theories; and poor feedback. In other domains in which these conditions do not hold, such as physics, chemistry, chess, and so on, Camerer and Johnson pointed out that experts' configural rules should be accurate and better than linear models. Again, from a Social Judgment Theory perspective, one must look at both sides of the lens model:

the task as well as the person. This discussion is not meant to imply that experts do not use nonconfigural rules, or that they are not sensitive to task characteristics. Substantial research indicates that they do—and are—depending on task characteristics. In a study of particular importance from a cognitive engineering perspective, Hammond et al. (1987) found that nine different information displays significantly affected the cognitive processes of expert highway engineers. To quote:

> Individual analyses of each expert's performance over the nine conditions showed that the location of the task on the task index induced cognition to be located at the corresponding region on the cognitive continuum index. Surprisingly, intuitive and quasi-rational cognition frequently outperformed analytical cognition in terms of the empirical accuracy of judgments. Judgmental accuracy was related to the degree of correspondence between the type of task (intuition inducing versus analysis inducing) and the type of the experts' cognitive activity (intuition versus analysis) on the cognitive continuum. (p. 753)

As the study by Hammond et al. (1987) showed, experts have a rich repertoire of mental representations and cognitive processes; much richer than those of novices. Whether the correct representations and processes are induced depends on task characteristics. The goal of cognitive engineering is, of course, to induce the correct mental representations and processes. Even so, we must remember that some tasks simply have a higher level of predictability than others. These are the tasks for which experts will perform best, regardless of the quality of the cognitive engineering. Such a position is essential in bounding the field's expectations and promises. Human judgment (and decision making) will always be an activity in which one will want perfect information (more accurately, perfect predictability), regardless of how well the system has been engineered cognitively. Our goal is to engineer the system so that people get the most that they can from the information that they have.

This concludes our brief overview of cognitive science research on human judgment. We now consider research findings on decision processes. We again caution the reader that our overview is small and selective. Our goal is not to survey all potentially applicable areas of research. Rather, we hope to commence the process of mapping computer technology to cognitive processes to enhance performance.

DECISION MAKING

This section covers six approaches to decision making: bounded rationality, prospect theory, decision strategies, Recognition-Primed Decision Making (RPD), Image Theory, and Social Judgment Theory (SJT) from the perspective of interpersonal learning and cognitive conflict. These approaches are descriptive in the sense that each one describes certain aspects of how people actually make decisions. They contrast with prescriptive approaches, such as decision analysis or other approaches based on economic models of decision behavior, that prescribe how one should make decisions, but do not necessarily describe decision behavior. Given our cognitive engineering focus, prescriptive approaches are not considered herein.

We begin this section by summarizing and extending some of the points made in the introductory section on judgment and decision processes. First, options are not generated in a vacuum. Rather, they are generated in response to our hypotheses with regard to what is happening, why it is happening, and its implications. Yet, not all approaches to studying decision making incorporate the decision makers' hypotheses about the situation, that is, the Hypothesis component in Wohl's (1981) SHOR paradigm. This is particularly the case for approaches derived from economic models of rational decision making, such as bounded rationality and prospect theory. For example, although bounded rationality explicitly incorporates various "states of nature," these states typically represent future events and not causes of the current decision problem. Prospect theory does not consider the Hypothesis component of the SHOR paradigm and, in fact, emphasizes option evaluation and selection, not option generation. Other approaches, such as Recognition-Primed Decision Making (RPD), not only incorporate but emphasize the importance of the situation when considering decision making.

Second, decision making is difficult, in part, because of two principal types of uncertainty: information-input uncertainty regarding one's judgments, and consequence-of-action uncertainty regarding the effects of different decision alternatives. Approaches derived from economic models of decision making emphasize consequence-of-action uncertainty. In contrast, RPD emphasizes information-input uncertainty, although it also considers consequence-of-action. Other approaches, such as Image Theory or the SHOR paradigm, address both types of uncertainty without emphasizing either one. Still other research, principally that focusing on different decision rules for combining information, only considers uncertainty in a few cases.

Third, all approaches emphasize the framing of the decision problem (Russo & Schoemaker, 1989). For some approaches, in particular bounded rationality, the problem is framed principally by the decision maker's cognitive ability to consider only a limited number of alternatives, evaluation criteria, and states of nature. In most approaches, however, the decision problem is framed by the interaction between the problem's representation (or wording) and the person's cognitive processes. This is emphasized in prospect theory in which framing the same problem in terms of gains or losses results in different editing and evaluation processes and, in turn, different choices. To quote Beach (1990, p. 23) when discussing frames in

Image Theory, "the frame is a schema that defines and that is defined by the image constituents that comprise it."

Fourth, all approaches emphasize the valuative nature of decision making. The approaches derived from economic models of decision making, as well as Social Judgment Theory and Image Theory, have specific means for representing these values in terms of both utility (or value) functions and relative importance weights (or tradeoffs) among competing evaluation criteria when decision makers chose among alternatives. In addition, Image Theory uses the concept of a value image to incorporate more higher order principles as guiding decision behavior.

Fifth, all approaches note that decision making is difficult in part because people have different values, goals, and images. In addition, individuals vary in their degree of insight into their own values. Among the descriptive approaches considered herein, only Social Judgment Theory employs specific, computer-based procedures for externalizing cognitive conflict among individuals. These procedures for providing interpersonal, cognitive feedback have considerable implications for the cognitive engineering of group decision support systems; therefore, they are considered herein.

Bounded Rationality

The concept of bounded rationality is attributed to Nobel laureate Herbert Simon (e.g., see 1955, 1979; see also Hogarth, 1987, and March, 1978, for general discussions), who argued that humans lack both the knowledge and computational skill required to make decisions in a manner compatible with economic notions of rational behavior. In order to deal with human and task limitations or bounds, Simon argued that humans simplified decision problems so that they can address them in a "reasonable" if not economically "rational" manner. In particular, Simon's approach was to specify what the rational economic model required humans to know and do, and then to ask how they could cope with the task given their limited knowledge, memory, and processing capabilities. Therefore, the approach is conceptually similar to that described earlier for cognitive heuristics research.

Task Characteristics. The rational model's requirements are illustrated by the concept of a payoff matrix, an example of which is presented in Table 2.1. The rows of the matrix represent all of the different alternatives (i.e., options) available to the decision maker for solving a particular decision problem. The columns represent all of the different states of the world, as defined by future events, that could affect the attractiveness of the alternatives. The $p_1 \ldots p_k$ values represent the probabilities for each state of the world. The cell entries in the matrix indicate the value (or "utility") of the outcome or "payoff" for each combination of alternatives and states of the world. Each outcome is presumed to represent a cumulative payoff comprised of perceived

TABLE 2.1
The Rational Economic Model's Decision Making Requirements
as Represented in a Payoff Matrix

	States of Nature			
Alternatives	$(p_1)S_1$	$(p_2)S_2$	. . . . .	$(p_k)S_k$
A	a_1	a_2	. . . . .	a_k
B	b_1	b_2	. . . . .	b_k
.	.	.	. . . . .	.
.	.	.	. . . . .	.
.	.	.	. . . . .	.
N	n_1	n_2	. . . . .	n_k

advantages and disadvantages on multiple criteria of varying importance to the decision maker. Finally, the rational decision maker is required to select the alternative that maximizes expected utility, which is calculated for each alternative by first multiplying the values for the outcomes and the probabilities for the states of nature and then summing the products.

Important Findings. The rational economic model clearly assumes that the decision maker has extensive knowledge and impressive unaided, computational power. In addition to Simon's research, substantial psychological research (e.g., see the reviews by Einhorn & Hogarth, 1981; Slovic, Fischhoff, & Lichtenstein, 1977) indicates the inadequacy of these (and other) assumptions of the model. Therefore, how do we cope with the cognitive demands represented in the decision matrix? How does unaided human decision behavior remain purposeful and "reasonable" given the dynamic nature of the environment and inherent information acquisition and processing limitations?

Simon suggested three simplification strategies. First, people simplify the decision problem by only considering a small number of alternatives and states of nature at a time. Second, people simplify the evaluation problem by setting aspiration (or acceptability) levels on the outcomes. And third, people simplify the selection problem by choosing the first alternative that satisfies the aspiration level. In other words, people do not optimize (i.e., choose the best of all possible alternatives), but satisfice (i.e., choose the first satisfactory alternative). In this way, people can reduce information acquisition and processing demands and still act in a purposeful, reasonable manner.

The strategies in Simon's theory of bounded rationality are not, however, without their costs. First, as Hogarth (1987) has pointed out, research on creativity suggests that one of the biggest deficiencies in human decision behavior is one's failure to sufficiently imagine the range of alternatives at one's disposal and the various events that could occur in the future. Second, aspiration levels may be unrealistically high or low. The former could well result not only in the elimination of potentially good alternatives early in the decision process, but the acceptance of a relatively inferior alternative later in the process because subsequent events have forced one to lower his or her aspiration level. In contrast, unrealistically low aspiration levels and the satisficing strategy may well result in the acceptance of relatively poor alternatives early in the decision process.

It is important to emphasize that bounded rationality represents a descriptive theory of human decision behavior. It does not specify how people should make decisions but, rather, presents a theoretical perspective on how people do make decisions given a complex, dynamic environment and limited information acquisition and processing capabilities. Moreover, subsequent research indicates that people are quite capable of using other, in some cases more, complex strategies than the three proposed by Simon. In contrast, the rational economic model is now typically seen as a prescriptive not descriptive theory of decision making. It is typically referred to as decision theory (or expected utility theory or subjective expective utility theory), and it provides an axiomatic basis for specifying how people should make decisions, as represented in the decision matrix, given that they accept certain logically defined principles of behavior. Moreover, analytical procedures called *decision analysis* and various support systems have been developed to help people implement decision theory. There are many books on decision theory and decision analysis. The texts by Brown, Kahr, and Peterson (1974), Huber (1980), Watson and Buede (1987), and von Winterfeldt and Edwards (1986) are good introductions for the interested reader.

Prospect Theory

Kahneman and Tversky (1979) proposed "prospect theory" as a descriptive theory of choice under uncertainty. Like Simon's bounded rationality, prospect theory is juxtaposed against expected utility theory. What is particularly important about prospect theory for cognitive engineering is that it distinguishes between two choices in the choice process. The first is called *editing*: its purpose is to simplify the presented information in the choice setting in order to enhance decision making. The second phase is called *evaluation*: its purpose is to analyze the edited choices (i.e., prospects) so that the decision maker can select the one with the highest personal value. What Kahneman and Tversky have shown is that the way the prospect information is presented to people significantly affects how they edit and evaluate it so that information that should result in the same choice from the perspective of expected utility theory actually results in different choices.

Task Characteristics. Subjects in the experiments used to generate prospect theory are faced with simple prospects (or choices) that have a correct answer based on expected utility theory. For example, the following prospect is taken from Kahneman and Tversky (1979):

Choice A: $4,000 with p = .8; $0 with p = .2 or
Choice B: $3,000 for sure; that is, p = 1.0.

The majority of participants will select Choice B. Yet, Choice A has the greater expected value, that is, $4,000 × .8 = 3200. Now, consider the following prospect:

Choice C: −$4,000 with p = .8; $0 with p = .2 or
Choice D: −$3,000 for sure; that is, p = 1.0.

The only change in the second prospect is that the sign has been reversed so that one is now considering losses, not gains. However, in this case, the majority of the subjects picked Choice C. That is, they would now be willing to take a gamble of losing $4,000 with a probability of .8, which has an expected value of losing $3,200, instead of taking a sure loss of $3,000. Again, they have selected the choice with the lower expected value. In addition, they have switched from the "sure thing" to preferring the gamble.

Numerous other examples have been used to demonstrate that people's choices are not always consistent with expected utility theory. More generally, Hogarth (1987) pointed out that, by and large, the research has been directed toward testing the general principles of decision theory, which assume that people express consistent beliefs in the form of predictive judgments and consistent preferences in the form of evaluative judgments. The principles regarding preferences include transitivity (if A is preferred to B and B to C, then A should be preferred to C), dominance (if alternative A is preferred to alternative B on all dimensions, then there should be no way that, in total, B should be preferred to A), and invariance (one's preference for two options should not be affected by the way one presents information about them). "Perhaps the most striking feature of these principles is that, whereas they are accepted as reasonable when stated in the abstract form, their implications are often violated in actual choices" (p. 109). Prospect theory provides some insights as to why this occurs.

Important Findings. Kahneman and Tversky (1979) hypothesized that the editing and evaluation phases that people use to make decisions under risk have distinct operations. For example, Kahneman and Tversky proposed six editing operations: coding, combination, segregation, cancellation, simplification, and dominance. With respect to coding, people perceive outcomes as gains or losses from a referent point rather than final states (e.g., of wealth). The current position is usually considered as the referent point. However, the location of the reference point and, in turn, the coding of outcomes as gains or losses, can be affected by how the prospects are formulated, as well as by the person's expectations. This coding is particularly important in framing decisions because, as the prior example indicates, people tend to be risk adverse when considering gains and risk seeking when considering losses, particularly if one of the prospects is certain. The research by Tversky and Kahneman (1979, 1982) and others (e.g., McNeil, Pauker, Sox, & Tversky, 1982; Payne, Laughhunn, & Crum, 1980) indicates that (a) reference points can be manipulated, and (b) losses loom much larger than gains.

The other five editing operations are used to simplify the choice. For example, a prospect can be simplified by combining the probabilities associated with the same outcomes. However, this can sometimes result in inappropriate problem representations because, as the cognitive heuristics research indicates, people do not always implement probability theory correctly. Similarly, people appear to segregate the risky from the riskless components of prospects (called *segregation*) and cancel out aspects shared in common between prospects (called *cancellation*). Although quite reasonable operations for simplifying complex problems, they can result in different choices simply depending on how the problem is framed (e.g., see Hogarth, 1987).

Tversky and Kahneman proposed two operations in the evaluation phase of prospect theory: (a) a value function, which is analogous to the utility function in expected utility theory; and (b) a decision weight function, which indicates the subjective importance of the probabilities in prospects. The value function codes the psychological value of gains and losses from the "coded" reference point. The function is steeper for losses than for gains, consistent with the observation that losses loom much larger than gains. We tried to capture this with the reversal illustrated in the previous example. Moreover, outcomes near the reference point are given more value per unit change than units farther from the reference point. For example, assuming a reference point of $0 and a range from $0 to $1,600, a $1 difference between $0 and $10 typically has more psychological value than a $1 difference between $1,550 and $1,560.

The decision weight function in prospect theory links the probabilities in prospects to choice. In particular, the function represents the finding that people seem to overweigh low probabilities and underweigh high probabilities when compared to expected utility theory. Interestingly, prospect theory only defines the decision weight function between probabilities of 0 and 1. These two probabilities are given weights of 0 and 1, respectively, indicating the special effect of certainty. We also tried to capture this in our example.

Decision Strategies

Substantial research has focused on describing the different types of strategies people use to combine information into decisions. Extensive reviews of this research can be found in Beach (1990), Hogarth (1987), and Zsambok et al. (1992). In general, the literature makes a distinction between two classes of decision strategies: compensatory and noncompensatory. These different classes also have been referred to as "conflict-confronting and conflict avoiding" by Hogarth (1987, p. 72) depending on whether the strategy confronts or avoids inherent tradeoffs in the situation. To quote Hogarth:

> Conflict-confronting strategies are compensatory. That is, they allow you to trade off a low value on one dimension against a high value on another. For example, if you used a compensatory strategy, you would be prepared to balance the low job security of alternative A against its high career prospects and pay. Conflict avoiding strategies, on the other hand, are non-compensatory. That is, they do not allow trade-offs. For instance, you might decide that job security was very important to you and that you would not consider any position that was low on that dimension, no matter how attractive it was on other aspects. Thus, you would eliminate job A. (p. 92)

Most decision strategies do not consider probabilities. The two strategies that do consider probabilities—maximization of expected value and maximization of subjective expected utility—are represented conceptually by the payoff matrix presented earlier. For the maximization of expected value, one uses (a) objective probabilities to represent the probabilities for the states of nature, and (b) dollar values for the cell entries. The goal is to select the alternative that has the highest sum of the products for the probabilities and values across the different states of nature. This is the alternative that statistically has the highest expectation of maximizing one's payoff. For the maximization of the expected utility strategy, the objective probabilities are replaced by subjective probabilities and the dollar values are replaced by subjective utilities, which may be a weighted composite over a number of different dimensions. The selection strategy, however, remains unchanged. The decision maker should select the alternative with the highest expected utility, that is, the alternative with the highest sum of products across the states of nature.

The principal decision strategies in Hogarth (1987) and Zsambok et al. (1992) are listed next. We begin by describing the task characteristics used in research investigating these strategies.

Task Characteristics. Figure 2.8 presents the task representation used in the study of decision strategies. The representation is a decision alternative x evaluation dimension matrix. The rows are the alternatives, the columns are the dimensions the decision maker uses when evaluating the alternatives, and the cell entries indicate the alternative's value on the dimensions. For example, in the job selection case considered by Hogarth in the earlier quote, alternative A has low job security, but high career prospects and high pay. The best alternative among the set under consideration depends on the decision strategy.

The task represented in Figure 2.8 has been operationalized in a number of different ways. Two ways are considered here. The first way is by presenting the decision maker with individual cases, in which each case represents a decision alternative defined in terms of its values on the dimensions (or evaluation criteria). Decision makers simply rate the desirability of each

	Dimensions			
Alternatives	X_1	X_{2S}	$\cdots\cdots$	X_k
A	X_{a1}	X_{a2}	$\cdots\cdots$	X_{ak}
B	X_{b1}	X_{b2}	$\cdots\cdots$	X_{bk}
$\cdot$	$\cdot$	$\cdot$	$\cdots\cdots$	$\cdot$
$\cdot$	$\cdot$	$\cdot$	$\cdots\cdots$	$\cdot$
$\cdot$	$\cdot$	$\cdot$	$\cdots\cdots$	$\cdot$
N	X_{n1}	X_{n2}	$\cdots\cdots$	X_{nk}

FIG. 2.8. Alternative X dimension representation of task for studying decision strategies. From Hogarth (1987). Reprinted with permission.

case. This approach has been used within SJT and has been implemented on both time sharing and personal computers (PC) via the POLICY software package (Rohrbaugh, 1986). Multiple regression analysis typically has been used to analyze the judgments for the n number of cases. The judgments are the values on the dependent variable; the values on the dimensions are the values on the independent variables. Multiple regression analysis will result in a linear, additive model to represent the person's decision strategy unless other decision strategies are represented as quantitative models for predicting the person's judgments. The POLICY program has this capability.

The second approach is the information board developed by Payne (1976). The information board looks like the alternative × dimension matrix shown in Figure 2.8. When the session starts, the cell entries are covered; then the decision makers sequentially uncover the cell entries for which they want information. By observing the order in which the cells are uncovered, and listening to the decision makers think aloud, one can ascertain the type of strategies being used to reach a decision. The PC version of the information board has been called the mouselab system because a mouse is used to uncover the cells (Johnson et al., 1989).

Important Findings. This section lists the different types of compensatory and noncompensatory strategies observed in various research studies. Three compensatory models are considered first.

1. Linear additive strategy—the value of an alternative is equal to the sum of the products, over all the dimensions, of the relative weight times the scale value for the dimension.

2. The additive difference strategy—the decision maker evaluates the differences between the alternatives on a dimension by dimension basis, and then sums the weighted differences in order to identify the alternative with the highest value overall.

3. The ideal point strategy—is similar to the additive difference model, except the decision maker compares the alternatives against an ideal alternative instead of each other.

We now consider seven types of noncompensatory models:

4. Dominance strategy—select the alternative that is at least as attractive as the other alternatives on all the dimensions, but is better than them on at least one dimension.

5. Conjunctive strategy—select the alternative that best passes some critical threshold on all dimensions. This is the satisficing strategy when one selects the first option that passes a threshold on all dimensions. The conjunctive strategy is often used to reduce the set of alternatives by eliminating all alternatives that fail to pass a threshold on all dimensions.

6. Lexicographic strategy—select the alternative that is best on the most important dimension. If two or more alternatives are tied, select among them by choosing the alternative that is best on the second most important dimension, and so on.

7. Disjunctive strategy—select an alternative that passes some critical level on an important dimension. It can be used when no one alternative passes the threshold on the most important dimension or when a set of alternatives must be selected in which the alternatives have to be strong on different dimensions, such as an athletic team.

8. Single feature difference—find the dimension on which the alternatives differ most, and choose the alternative that is best on this dimension irrespective of the other features.

9. Elimination by aspects—sequentially identify different dimensions, either according to their importance or some more probabilistic scheme. Eliminate all alternatives that fail to pass the threshold or aspect for each dimension until only one alternative is left.

10. Compatibility test—compare alternatives to a threshold on each dimension; for each alternative, compute the (weight) sum of the dimensions for which the alternative fails to meet the threshold; reject all alternatives whose sum is larger than a selected critical value; if only one alternative remains, select it; otherwise, choose the best alternative by using some other strategy. The compatibility test is an important aspect of Image Theory and is considered further later.

In closing this section, we note two general findings. First, research employing the information board has shown that decision makers often use multiple strategies when considering decision alternatives. Typically, they use noncompensatory strategies to reduce the number of alternatives and dimensions under consideration. To stay with the job selection example, a person might first eliminate all alternatives that fail to pass a specific threshold on security, which may no longer be as important when considering the reduced set of alternatives. Then, after the set of alternatives and dimensions have been reduced to a smaller, more manageable set, people often employ a compensatory strategy in which they weigh the strengths and weaknesses of the alternatives in order to select the one that best satisfies their values.

Second, the linear additive model has been shown to be extremely accurate in predicting people's desirability ratings. However, Hogarth (1987, p. 74) summarized the position of a number of researchers by concluding that "as a description of choice processes, the linear model is often inadequate. It implies a process of explicit calculations and the trading off of dimensions which, when there are many alternatives and dimensions, is not feasible for unaided judgment." The predictive power of the linear model is no doubt due, in part, to its robustness, that is, its insensitivity to deviations from underlying assumptions that permit it to reproduce the judgments actually generated by another process with a relatively high degree of accuracy. In addition, however, Hammond (1988) argued that it and other function-relation strategies are induced by many tasks. The predictive accuracy of the linear model, its ease of implementation in computer systems, and the fact that many tasks induce function-relation-seeking strategies make the linear model on appealing strategy from a cognitive engineering perspective, as is shown in the section on cognitive conflict. However, one should consider alternative strategies if it is critical that the explicit decision strategy be captured in a particular application.

Recognition-Primed Decision Making

As noted in the introductory section, research by Klein and his associates (e.g., Klein, 1989, 1993; Zsambok et al., 1992) has shown a strong relation between judgments of the situation and decision making, particularly among experienced decision makers. In fact, the relation has been so strong that Klein referred to his model of decision making as a Recognition-Primed Decision (RPD) model.

A pictorial representation of the RPD model is presented in Figure 2.2. It shows the proficient decision maker becoming aware of events that have occurred and relying on experience to recognize these events as largely typical. The judgments of typicality includes guidance for plausible goals, critical cues, expectancies, and typical actions. The simplest case is one in which the situation is recognized and the obvious reaction is implemented.

A somewhat more complex case is one in which the decision maker performs some conscious evaluation of the reaction, typically using imagery to uncover problems. The most complex case is one in which the evaluation reveals flaws requiring modifications or the action is judged inadequate and is rejected in favor of another typical reaction. In all cases, however, decision making is guided by the recognition (or understanding) of the situation. The verification/nonverification of expectancies can alert the decision maker that the situation understanding is wrong and that it is time to rethink it and to gather more information before preceding with decision making.

Klein's RPD model has different levels of "recognition" and, therefore, delineates different cases under which decision makers generate and evaluate individual or multiple alternatives. The different levels are determined by the decision maker's experience and task characteristics, such as time pressure, situational uncertainty, and the degree to which one's decision have to be justified after the fact. These different levels are described in detail later under important findings. First, however, we describe the task conditions typifying RPD research.

Task Characteristics. As discussed briefly within the context of expertise research, RPD research has been conducted primarily in naturalistic versus laboratory settings. By *naturalistic* is meant the decision maker's actual, real-world setting. As Orasanu and Connolly (1993) pointed out, such settings are epitomized by eight factors: the problems are ill structured; they occur in environments with incomplete and imperfect information; the decision maker's goals are often shifting, ill-defined, and competing; there exist action/feedback loops of various degrees of quality; decisions are typically made under significant time pressure; the stakes are high; there are multiple players; and, because most decisions are made in organizational contexts, the decision maker must address organizational goals and norms.

Klein and his associates developed the Critical Decision Method (CDM) as a means for understanding how people make decisions in naturalistic settings. CDM is a retrospective interview strategy that applies a set of cognitive probes to actual nonroutine incidents that require expert judgment or decision making. Once the incident is selected, the interviewer asks for a brief description. Then a semistructured format is used to probe different aspects of the decision-making process. These aspects include, for example, the cues and knowledge used in the situation, previous experiences (i.e., analogues) that came to mind, and the goals, alternatives, and reasoning involved in the decision process. Klein et al. (1989) have obtained high levels of interrater agreement for coded information taken from transcripts over a wide range of domains. Moreover, they have obtained comparable results using more traditional laboratory tasks, thereby supporting the viability of the CDM approach.

Concept maps can be used to represent the result of CDM interviews pictorially. Concept maps are graphs that identify what concepts appear to be linked together cognitively. The concepts are objects or events that are arranged in a bubble-like diagram, in which each node (or bubble) represents one concept. The nodes are linked together with directional arrows that specify the nature of the relationship. Concept maps often have a hierarchical structure going from the more general concepts at the top of the map to the more specific concepts at the bottom of the map. This structure has been found to be quite useful in providing a quick and flexible schematic summary of the domain of interest. For illustrative purposes, Figure 2.9 from Wolf, Klein, and Thordsen (1991), presents a simplified concept map for driving a car.

To quote Wolf et al. (1991, p. 3) as to the function of concept maps:

> First, they are an overt, explicit representation of individual concepts and the linkages among them. The structure of concept maps can be likened to a semantic net of the task. Second, they allow knowledge engineers and SMEs [subject matter experts] to exchange views and correct misunderstandings as the map is being developed. In addition, a completed map provides an explicit record of the conceptual path followed during development. When misunderstandings do occur, it is very easy to trace back through the map to rectify the problem. Finally, a completed map can be viewed in terms of how different concepts are clustered together. By examining the map in this way, the designer can obtain a perspective of the 'big picture' of the task—in other words, determining what pieces of information, cues, and goals allow the decision-maker to assess different situations.

Important Findings. The most important finding of RPD research is the essential nature of decision makers' recognition of what is happening in the environment and, thus, what actions are best for addressing it. It is

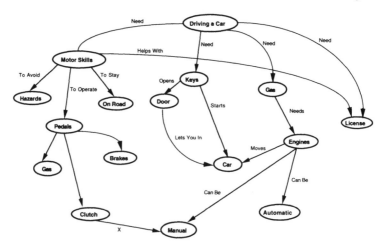

FIG. 2.9. Concept map for automobile operation.

this explicit linkage between judgment and decision that differentiates the RPD model from, for example, the three approaches to decision making considered thus far. Therefore, this is what we emphasize in the brief research summary presented here.

Figure 2.10 from Zsambok et al. (1992), presents a pictorial representation of a decision event. A decision event occurs when a decision maker identifies that a problem exists and that a course of action (COA) is needed or desired. The decision maker first attempts to diagnose the situation, that is, find out what the problem is, understand what caused it, and assess its potential consequences. The research on juror decision making by Pennington and Hastie (1986) suggests that situation diagnosis typically involves generating a story to explain the data. The story not only makes sense of the situation, it frames it so that the decision maker knows what types of actions are most appropriate for resolving the situation.

Diagnosis may be immediate based on the decision maker's understanding of what the available information means. In such cases, the decision maker recognizes the situation based on the familiarity of its features and the explanation for them. For other situations, however, accurate diagnosis may require that additional information be collected and, therefore, may be prolonged for some period. In addition, as the event unfolds over time, the decision maker may attempt to "manage the situation" (p. 34), that is, the decision maker may attempt to alter the course of the event while gathering information for diagnosis. These management activities occur prior to making a decision that implements a specific course of action.

Zsambok et al. (1992, p. 52) identified three types of management decision strategies. The first type are incremental decisions, descriptively referred to as "hedgeclipping" by Connolly and Wagner (1988). This refers "to the

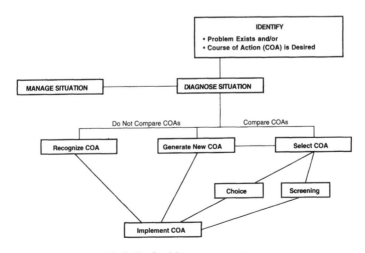

FIG. 2.10. Decision event structure.

special case where the decision-maker does not expect to diagnose the situation in order to arrive at a decision about an appropriate COA. Rather, the decision-maker expects to continuously shape the event until it is resolved satisfactorially" (Zsambok et al., 1992, p. 52).

The second type of management decision is planning how to shape the course of events. The focus here is not on planning how to best implement the chosen COA, but rather on planning how to manage the situation. And the third type of management decision strategy is managing the decision-making process. This has been called *metacognition* (e.g., see Cohen & Hull, 1990; Glaser & Chi, 1988). "It involves monitoring, or reflecting on how decisions are being made, knowing about resources available to use in decision-making, and modifying the decision-making process where necessary to make better use of the available decision-making resources" (Zsambok et al., 1992, p. 53). Russo and Schoemaker (1989) used the term *metadecision* to describe the process of deciding how to decide at the very beginning of the decision-making process.

After the situation is diagnosed and/or managed, the decision maker makes a decision, that is, arrives at a COA as in Figure 2.10. RPD distinguishes three ways to do this: through COA recognition, generation, or selection. *Recognition* refers to situations that are so familiar to decision makers that they can recognize what to do from memory. *Generation* refers to situations in which decision makers have to generate the COA either because the situation is not familiar or because it is familiar, but the last COA failed to resolve it.

Thus far it has been assumed that there is only one COA. *Selection* refers to those situations in which the decision maker needs to choose from a set of alternative COAs. This may occur either because the decision maker or other players generated more than one COA or because they were available in the situation. Regardless, selection requires a comparative evaluation of COAs. As Figure 2.10 suggests, the selection process may include the initial screening of alternative COAs, for example, by using some of the noncompensatory decision strategies considered earlier, as well as the actual choice of the COA. Finally, the decision event is terminated when the decision maker implements a COA.

Figure 2.11 presents a summary of the decision strategies used during situation diagnosis and for arriving at a COA.

Image Theory

Beach's (1990, p. 3) "informal statement of image theory" emphasizes five concepts: images, framing, adoption, progress, and deliberations. Image theory views the decision maker as employing three types of images that partition one's decision-related knowledge. The first image is the *value image*. This image "defines how events should transpire in light of the decision-maker's values, morals, ethics, and so on" (p. 3). The second image is the

---

A. During Situation Diagnosis:

    1. Feature Matching to Recognize the Situation

    2. Generating a Story, or an Explanation-Based Situation Assessment

    3. Managing

        a. The Event Via Incremental or Planning Decisions
        b. The Decision-Making Process Via Metacognitive Decisions

B. For Arriving at a Course of Action:

    1. Recognizing a COA as a Result of Situation Diagnosis

    2. Modifying a Recognized COA (Non-Comparative Evaluation)

    3. Selecting a COA (Comparative Evaluation)

    4. Generating a New COA

FIG. 2.11. Decision strategies used during situation diagnosis and for arriving at a course of action.

trajectory image. This image is basically "an agenda of goals and related time-lines for accomplishing them" (p. 3). And the third image is the strategic image. This image includes (a) one's general plans and specific actions (called *tactics* by Beach) for goal accomplishment, (b) one's projections about how well these plans and actions will accomplish relevant goals, and (c) after the plans and actions are implemented, how well they are working toward goal accomplishment. This last component of the strategic image is referred to as a *progress decision.* It is particularly important to situation assessment judgments and is considered in more detail later.

For illustrative purposes, we present the description of the three images in terms of Beach's (1990) image theory analysis of auditor decision making. Specifically, the value image contains the general ideas of what constitutes (a) acceptable business and accounting principles, and (b) acceptable auditing standards and techniques, including appropriate compliance and substantive tests. The trajectory image consists of the goal of a correct audit for the client(s), "where a correct audit is defined as whether the audit process supports issuing an unqualified opinion (i.e. lack of material error in the client's financial statements)" (p. 201). And the strategic image consists of the audit plans for examining the client's financial statements, the forecasted predictions the auditor uses to assess the audit's progress toward accomplishing the goal of a correct audit, and the results of tests of these forecasts against the actual data.

Aspects of the images are not relevant in every context. What makes them relevant is a process Beach (1990, p. 8) called *framing*: "In order to interpret events and to bring relevant knowledge to bear upon them, the decision-maker relies upon recognition or identification of the present context to define a subset of the constituents from his or her images as relevant to the decision at hand." When everything is progressing smoothly, as defined by the framed context, decision makers need only monitor the situation.

However, if things are not progressing smoothly, then they must intervene. Intervention may include the adoption of new actions or plans comprising the strategic image and even new goals for the trajectory image. The latter case is illustrated in auditor decision making when the auditor concludes that the data do not support the goal of accomplishing a correct audit, that is, issuing an unqualified opinion.

There are two kinds of decisions in Image Theory: adoption decisions and progress decisions. Adoption decisions focus on which values, goals, plans, and tactics (i.e., actions for implementing the plans) should be constituents of the value, trajectory, and strategic images, respectively. Progress decisions focus on whether a particular plan (or tactic) on the strategic image is producing satisfactory progress toward goal attainment on the trajectory image. Accurate situation assessment is critical to making good progress decisions. For if the situation has changed in ways that would cause the current plan and actions to fail in achieving one's goals, the earlier this judgment is made, the higher will be the probability of successful recovery via new plans and actions.

An important aspect of Image Theory is the tests used to make adoption and progress decisions:

> The compatibility test assesses whether the features of a candidate for adoption 'violate' (are incompatible with) the relevant (framed) constituents of the various images, and whether the forecasts based upon the constituents of the strategic image 'violate' the relevant constituents of the trajectory image. The profitability test, which applies only to adoption decisions, assesses the relative ability of competing candidates to further the implementation of ongoing plans, attain existing goals, and comply with the decision-maker's principles. The object of the compatibility test is to screen out the unacceptable. The object of the profitability test is to seek the best. (Beach, 1990, p. 9)

The last concept is deliberation. Beach pointed out that most decisions are straightforward and, therefore, require little deliberation. Other decisions are, however, not so easy. This occurs when the screening process does not generate a clear candidate for adoption. In such cases, deliberation is required in order to clarify relevant principles, goals, plans, and actions to various degrees. It also permits the decision maker to forecast the possible effects of various plans and actions under different conditions. Deliberation is the essence of decision making and brings together the other four concepts.

Task Characteristics. Image Theory integrates three lines of research. The first line is decision research. Many of the initial studies performed by Beach, Mitchell, and their students were efforts to apply and extend decision theoretic concepts to more adequately address human decision behavior. The more recent studies have tested and applied Image Theory, which represents a rejection of decision theory. The second line of research that Image Theory incorporates is cognitive science research on knowledge representation, planning, and framing. Most of this research uses traditional experimental designs, as was covered in the sections on expert versus novice problem solving and prospect theory. The third line of research is organizational science. This research (e.g., Mintzberg, 1979), like RPD research, has emphasized the use of interviews with decision makers and the observation of actual decision-making groups.

Important Findings. Image Theory has made a number of important findings. Perhaps its most important is its conclusion, based on substantial empirical research, that a decision theoretic framework is an inadequate representation of decision processes. We quote Beach (1990, pp. 170–171) on this point, but have omitted his many references:

> Over the last 15 years my colleagues and I have examined a broad range of real-life decisions that were extremely important to the participants in our studies. . . . One of the first things we learned while doing these studies is that probabilities mean little to decision-makers and have surprising little impact on their decision. Moreover even if they did, decision-makers' probabilities bear little relationship to decision theorists' probabilities. Risk is a factor, yes, but it is viewed as a global, negative aspect of the candidate rather than as a weight on each of the candidate's potential outcomes. Probability is of little concern because decision makers assume that their efforts to implement their decisions will be aimed, in large part, at making them happen. This assumption makes probability, in any random sense, irrelevant. . . .
>
> The second thing we learned is that most important decisions are compatibility decisions. They involved one candidate, a single alternative to the status quo. In these decisions, the status quo is seldom viewed as a competing alternative—it is the baseline against which possible change is evaluated. . . .
>
> The third thing we learned is that even compatibility decisions can be forced into an expected value—or, as was the case in most of our studies, multi-attribute utility analysis. Although it was inappropriate, imposing this structure on the decisions often proved enlightening both to our participants and to us because it made them do the main thing that decision aiding ought to do: it made them think in a systematic way about the various ramifications of the decision and how they felt about them. . . . However, . . . it is not the best way to go about things. . . . [Decision] aids should be congruent with the ways that decision makers think about their decisions.

Because of its integrative nature, Image Theory addresses aspects of each of the approaches to decision making considered here. For example, it incorporates Simon's (1955) concepts of bounded rationality in terms of how the decision is framed and the concept of satisficing in terms of the compatibility test. However, unlike Simon's concept of satisficing, the compatibility test can produce two (or more) candidates. Then, the profitability test is employed to select the best one.

Second, Image Theory emphasizes the concept of framing found in Kahneman and Tversky's (1979) prospect theory, although not necessarily its value and decision weighing functions. Third, Image Theory argues that, in contrast to the compatibility test, profitability tests can be made in many ways—"from flipping a coin, to asking an authority what to do, to decision analysis" (Beach, 1990, p. 127). In particular, Image Theory considers the different compensatory and noncompensatory decision strategies considered here as potential strategies that decision makers might use for the profitability test based on their knowledge of these strategies and characteristics of the environment. A cost-benefit, subjective-expected utility theory logic is proposed (and empirical data provided) as a metadecision process for actually selecting strategies for the test.

Fourth, Image Theory and the Recognition-Primed Decision (RPD) model share a number of similarities. For example, both emphasize the linkage between judgment and decision making. In particular, Image Theory also incorporates the concept of recognition-primed decision processes when the present context is virtually identical to "contextual memory" (Beach, 1990, p. 51) and the notion of option generation when the situation is identified as similar, but not identical to these memories. Both strategies emphasize the generation and refinement of individual options versus the systematic evaluation of multiple options, although both approaches can handle the latter case as well. Both approaches stress the importance of goals, cues, expectations, and forecasts for effective judgment and decision making. Both employ Pennington and Hastie's (1986) Story Model as a cognitive explanation for how people generate and execute mental simulations to make judgments about situational causes and to forecast action consequences. And both approaches emphasize the importance of considering environmental and person characteristics that can impair judgment and decision-making accuracy. However, unlike Social Judgment Theory, neither approach has yet attempted to model the effect of environmental characteristics on judgment or decision processes. Nor has either approach systematically incorporated the research on judgment heuristics and biases.

There also are differences between the two approaches. The RPD model does not employ the concept of images, at least not explicitly. Nor does it systematically utilize the concepts of framing, frames, schemas, or the various other knowledge representation structures described in the literature, although they are mentioned. In contrast, Image Theory does not systematically utilize the concept of situation management strategies—strategies designed to alter the event without being a final course of action—that are considered in recent descriptions of RPD (Zsambok et al., 1992). Such strategies could be articulated, however, as part of the strategic image. Zsambok et al.'s recent description of RPD discussed the concept of metacognition, which is the introspective process of examining one's judgmental process. Image Theory does not do so, but it does employ the concept of metadecision—which RPD does not—which is the introspective process of deciding how to decide (Russo & Schoemaker, 1989).

Finally, Image Theory acknowledges the importance of the environmental side of the lens model in Social Judgment Theory (SJT). To quote Beach (1990, p. 21), "images, schemata, knowledge partitions, or whatever you wish to call them, ultimately derive their conceptual and theoretical value from the assumption that they bridge the gap between the actor's internal representation of phenomena and the empirical validity of that representation." However, neither Image Theory nor any of the other approaches discussed here has yet attempted to model the effect of environmental characteristics on judgment or decision processes. (In contrast, SJT has not been presented in a way that captures the range of concepts found in RPD and Image Theory.) We now turn to consider SJT research from the perspective of cognitive conflict and interpersonal understanding.

Social Judgment Theory and Cognitive Conflict

Considerable SJT research has focused on the situation in which two or more persons work together to perform a task. Figure 2.12 shows the "triple system case" in which one system is the task and the other two systems are the cognitive systems of the people. As can be seen, the triple system case is a natural extension of the double system case shown in Figure 2.4. Moreover, the Lens Model Equation (LME) can be used to assess the task performance parameters of each person and, more importantly, the agreement between them. That is, r_a can be used to represent the overall level of agreement in the decisions (or judgments) of the two persons, G can be used to represent the similarity in their decision processes, and Rs_1 and Rs_2 can be used to represent the cognitive control of each person's process.

Consequently, low agreement in a cooperative decision-making task can be the result of dissimilar cognitive processes and/or low cognitive control. Dissimilar cognitive processes are typically the result of poor interpersonal understanding of how each person is combining information to make a judgment. This low level of understanding is caused by environmental characteristics, such as nonlinear functions, correlated cues, and low levels of task predictability. Research has shown that cognitive technology can enhance this understanding, as well as cognitive control, and hence agreement and subsequent performance. Before considering this finding in more detail, we overview the characteristics of SJT tasks studying interpersonal understanding and agreement.

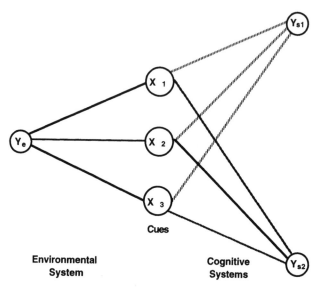

FIG. 2.12. "Triple System" case.

Task Characteristics. As was mentioned earlier in this review, cognitive technology has been used to improve understanding and communication in both the laboratory (e.g., see Hammond & Brehmer, 1973) and in real-world applications (e.g., see Hammond & Adelman, 1976; Hammond & Grassia, 1985). In this section, we first overview the characteristics of tasks used in laboratory research and then those used in real applications.

Brehmer (1976) and Rohrbaugh (1988) reviewed a host of laboratory studies evaluating the potential value of cognitive technology to improving understanding and reducing conflict. The basic approach is to first separately train members of the eventual decision-making group to have different strategies, called *policies*, for making inferences and/or decisions. For example, Person 1 may be trained to place a high relative weight and a negative linear function on Cue A, but no weight on Cue B when making judgments. In contrast, Person 2 may be trained to put no weight on Cue A, but a high relative weight (with an inverted U-shaped function form) on Cue B. After each person is trained, they are brought together to make group decisions.

In the group task, the task side of the lens model is structured so that each person must modify their policy by learning from the other person in order for the group to perform well. Continuing with this example, the group task may require that equal weight be placed on Cues A and B, that a negative linear function form be used to relate values on Cue A to levels of Desirability, and that an inverted U-shaped function form be used to related values on Cue B to Desirability. Control groups would have to learn to make group judgments by talking with each other and getting outcome feedback, that is, by the typical approach. The cognitive technology (or "cognitive feedback") group would be shown each person's judgment policy, that is, they would receive in both pictorial and textual form a description of how each person combines information to make his or her desirability judgments.

The approach is quite similar in real-world applications, except there is no need to train people to have a policy; they come with one. The first step is to uncover them. This is accomplished by having group members identify the cues that are important to the decision at hand. Once consensus is reached on the cues, each person is asked to evaluate a set of cases comprised of different values on the cues. These cases may be real ones (if there are not excessively high correlations between cues) or hypothetical ones that represent real cases if a sufficient number of the latter are not available. The values on the cues represent the values on the independent variables in a multiple regression equation. The desirability ratings represent the dependent variables. Each person's desirability ratings are regressed on the independent variable values in order to obtain a best-fitting model that represents his or her policy.

With the use of software designed for personal computers (Milter & Rohrbaugh, 1988; Rohrbaugh, 1986), each person can make the ratings online and be immediately shown the model representing his or her policy. If aspects of the model (e.g., relative weights, function forms, or combinational rules) are found unacceptable, changes can be made online. This new policy can then be applied to the cases so that the person can evaluate its implications. The initial stage is concluded when each group member has a model that each thinks accurately represents the decision policy.

The cognitive feedback stage can proceed in a number of different ways, but it typically follows the approach used in the laboratory research. That is, group members are first shown their policies so that they can compare their similarities and differences. Then, they are shown how the different policies result in different ratings for specific cases so that group members better understand the implications of the different positions. Third, the group members try to define a consensus strategy for making future decisions. This group policy is then applied to the previously evaluated (and sometimes new) cases so that the group members can evaluate its implications. Subsequent changes to the group policy can be made and the policy reapplied to the cases until a mutually acceptable position is reached.

It is important to emphasize that the SJT approach represents only one approach to modeling cognitive processes for the purposes of improving communication. In particular, decision analysis has been extensively used in a decision conferencing setting to facilitate group decision making (e.g., see Adelman, 1984; Weiss & Zwahlen, 1982). Decision analysis is comprised of a variety of different techniques for representing individuals' inferences and valuative judgments. Dating from its application by members of the

Commander-in-Chief of the European Command (EUCOM), who were faced with the decision of whether or not to evacuate U.S. nationals from Lebanon in 1976 (Kelly et al., 1981), decision analysis and the software used to support it have been effectively used to conduct decision conferences with decision-making groups for more than 15 years now.

More recently, group decision support system technology has been proposed (e.g., see Andriole, Ehrhart, Aiken, & Matyskiela, 1989; DeSanctis & Gallupe, 1987; Sage, 1991) as a more general approach to incorporating many different techniques for facilitating group decision making. In addition, artificial intelligence techniques have been used to capture expertise in the form of expert systems. However, one thing that SJT, decision analysis, group decision support systems, and artificial intelligence all have in common is the importance of providing information about cognitive processes (*cognitive feedback*) as a means of improving decision making. As such, they represent cognitive engineering approaches to improved group decision making.

Important Findings. SJT research has led the way in demonstrating the importance of cognitive feedback to improving interpersonal under-standing and group decision making. These findings are important to consider when designing computer technology to enhance group decision making. Reviews of the laboratory research can be found in Brehmer (1976) and Rohrbaugh (1988); a direct extension to negotiation and mediation can be found in Balke, Hammond, and Meyer (1972) and Mumpower (1988); and reviews of applications can be found in Adelman (1988), Hammond and Grassia (1985), and Hammond, Rohrbaugh, Mumpower, and Adelman (1977). Some of the important findings are considered, in turn.

First, interpersonal understanding is often made difficult simply by the nature of people's judgment policies. For example, people have great difficulty in learning another's policy if that person uses nonlinear function forms, complex rules for combining information, or has low cognitive control. In short, the same things that make it difficult for people to learn the complex inference tasks represented by the "environmental system" in the lens model also make it difficult for people to learn about and understand one another.

The second major finding is that discussion is an ineffective medium for interpersonal understanding. First, we have poor self-insight into how we make judgments and decisions. As Hammond (1976, p. 4) pointed out:

Verbal representations of introspective reports of covert mental operations are poor representations because of the well-known inaccuracy of introspection; that is, despite the best of intentions, any introspection regarding the basis for a judgment may simply be an incorrect description of the cognitive activity that took place ... the conceptual or technical ability required to describe their judgment processes completely; indeed they will not even know what a complete description should consist of.

Second, language, for all is beauty and versatility, is often an ambiguous medium for communication. As was noted in the review of research on cognitive biases, Kent (1964) and Moore (1977) showed that people attach very different meanings to the same expressions of uncertainty; Fischhoff et al. (1980) and others showed that the answer one gets depends on how one asks the question. Moreover, in addition to being ambiguous, Hammond and Boyle (1971, p. 106) noted that language is linear:

> Linear in the sense that it generally conveys relationships singly and in sequence. Thus, severe demands are made on the learner's ability to remember and to integrate sequentially presented relationships. Consequently, even if a verbal description of policy is accurate (which is unlikely, because of the quasi-rational nature of policy judgments; see Summers, Taliaferro & Fletcher, 1970) the listener is unlikely to be able to use effectively the information available to him (see Miller, Brehmer, & Hammond, [1971]).

Third, ambiguity in language is confounded by inconsistency in judgment. SJT research clearly shows that inconsistency in judgment (i.e., less than perfect cognitive control) is the typical state of affairs. That is, even if one can perfectly describe one's policy, one cannot perfectly implement it. Focusing on outcome feedback (i.e., the other's judgments or decisions) versus cognitive feedback (i.e., the other's policy) is an ineffective form of learning. Moreover, inconsistency between the decisions one makes and the descriptions one gives often causes confusion and sometimes distrust. The problem is potentially compounded by logical fallacies in recall and the fundamental attribution bias of focusing on the person rather than the situation.

The third major SJT finding is that cognitive feedback improves interpersonal understanding and group decision making. In particular, using computer technology to communicate cognitive processes addresses each of these three limitations with language. First, by presenting a model to represent one's policy one overcomes the lack of self-insight. One can explicitly show people the relative importance they placed on the different pieces of information, the function forms relating values on each cue to their judgments, the combination rule that appeared to best predict their judgments, as well as measures of cognitive control. Second, this information can be presented pictorially; it does not have to (although it can) be presented mathematically. Different sizes, colors, graphs, and so on, can be used to represent different cognitive processes. Third, these pictorial representations can be used to represent similarities and differences in the judgment policies of different persons.

The fourth major finding is that difficult decisions can often be resolved by first separating experts' judgments from value judgments; second, by having experts make the factual judgments and decision makers make the value judgments; third, by using cognitive feedback procedures both to

represent each group's judgment process and to quantify each one's judgments; and, fourth, by employing analytical methods for combining (quan-
tified) value and factual judgments so that decision alternatives can be evaluated systematically against a standard of values.

The conceptual linkage between the different types of judgments inherent
in decision making can be represented pictorially by the multilevel lens
model shown in Figure 2.13 from Adelman, Stewart, and Hammond (1975).
Again, there are two systems: the environment and the person. On the
left-hand side of the lens, the distal variable Ye is related to general theoretical
aspects of the environment (g1 and g2), which, in turn, are related prob-
abilistically to more observable cues—(A1 . . . An) and (B1 . . . Bn)—as in
the single-level lens model. On the right-hand side of Figure 2.13 are the
judgments and decision made by the person. That is, the person integrates
the cue information to make judgments (G1 and G2) about the general
abstract aspects of the environment (g1 and g2). Then, these judgments are
used to make an overall decision (Ys) about what to do.

The initial arithmetic formulation for separating and integrating factual
and valuative judgments is shown in Figure 2.14. The overall acceptability
of a decision alternative (Ys) is a function of the relative importance (Wi)
attached to the outcome dimensions (Gi). The relative importance judgments
should be the value judgments of decision makers; they should not be made
inadvertently by experts arguing for one alternative or another. In contrast,
an alternative's scores on the outcome dimensions (Gi) should be the factual
judgments of appropriate experts; they should not be made by decision

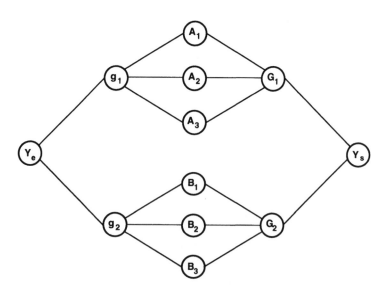

FIG. 2.13.   Multilevel Lens Model.

makers playing the role of experts. In short, value and factual judgments should be kept separate and distinct, so that each type of judgment is made by the appropriate person and charges of bias can be minimized.

It is important to note that when considering policy disputes employing social value judgments, there is no true Ye (i.e., correct answer) in reality. However, there are hierarchical judgment problems, such as in clinical inference (Hammond, Hursch, & Todd, 1964) or weather forecasting (e.g., Lusk & Hammond, 1991), in which there is either a true Ye or an agreed-on theoretical construct for representing it.

Value and factual judgments are recombined analytically by means of an equation. The initial and most frequently used equation is the additive equation shown in Figure 2.14, which is based on the multiple regression model used so predominantly by SJT researchers. Hammond et al. (1977) and Adelman (1988) pointed out, however, that other equations can be used to represent other organizing principles for combining factual and valuative judgments. The key point is that the final evaluations are made analytically by an equation, not intuitively in one's head. Consequently, SJT theorists argue that decisions are made in a consistent, systematic fashion using the same value standard; they are open to examination, thereby lessening charges of bias and vested interests. Moreover, using an equation facilitates sensitivity analyses that attempt to determine how sensitive the recommended alternative (that with the highest Ys) is to changes in value and/or factual judgments. Thus, the equation provides important cognitive feedback because it not only indicates which alternative's predicted outcomes best satisfy the value judgments guiding the decision, but how large the change in value or factual judgments must be to change the recommendation.

In closing this section it is important to emphasize the vast cognitive engineering potential that now exists for developing computer systems that

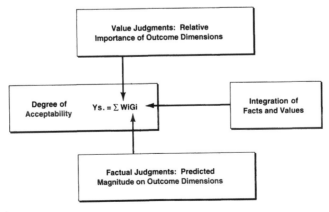

FIG. 2.14. Model for integrating and separating facts and values.

can enhance the interpersonal understanding of cognitive processes and, thereby, facilitate decision making. Our understanding of how people make decisions and the analytical methods for supporting them have expanded greatly within the last 20 years. At the same time, the capabilities of computer technology have, it seems, exploded, whereas the costs have decreased. These capabilities now include hypermedia, multimedia, simulation, animation, cost-effective color, sound, multidimensionality, and true multitasking, among others. The task before us is now to learn how to tailor the use of these capabilities so as to enhance how people think, communicate, decide, and act.

MEMORY AND ATTENTION

The next two sections briefly review the cognitive research literature on memory and attention, in turn. As we indicated in the introduction, our review was meant to be selective, not exhaustive. In particular, our goal has been to highlight the research on judgment and decision processes, for they represent the higher order cognitive processes that utilize memory and attention skills for effective problem solving. This is not meant to imply that memory and attention are not important. Obviously they are, for one cannot make good judgments and decisions without (a) remembering critical information and accurate relations between events and objects, or (b) attending to critical aspects of the environment, particularly under high workload conditions. Rather, our principal focus has been on the higher order processes they support.

Cognitive Science Findings: Memory

We begin this discussion by distinguishing between short-term memory and long-term memory. *Short-term memory* refers to the memory for information that has just been received and on which processing operations are still being performed. *Long-term memory* refers to the repository of knowledge and corresponds to what most people call memory.

We must also distinguish between the three operations comprising memory: encoding, storage, and retrieval. *Encoding* refers to the classification of sensory information that occurs during short-term memory. "Usually only the essence of what was sensed will be encoded" (Bailey, 1989, p. 139). *Storage* refers to the association of encoded information, that is, to the network of associations found in long-term memory. Finally, *retrieval* refers to the search process that obtains knowledge stored in long-term memory.

Forgetting may be due to a failure in any three of these operations. That is, information may not get encoded in the first place; consequently, it is not really there (in long-term memory) to be forgotten. Or, information may simply get lost (e.g., like a sieve) over time; consequently, it is no longer in long-term memory to be retrieved. Also, the information may not be retrievable; consequently, it may be in long-term memory, but one cannot access it.

Three findings regarding short-term memory (i.e., encoding) are now presented. First, "the short-term memory store is small and can hold about six units of information (sometimes referred to as chunks of information. . . . People forget longer messages sooner. Therefore, when working with short-term memory, the shorter the code, the better" (Bailey, 1989, pp. 134–135).

Second, what is perceived and, therefore, encoded into short-term memory is affected by one's expectations, that is, the associations maintained in long-term memory. Perception, and therefore memory, is not comprehensive but selective and limited. Therefore, "to remember any new piece of information, it must be associated with something the person already knows" (p. 146). Information coded along multiple sensory dimensions is less likely to be forgotten.

Third, one can use rehearsing and mnemonics to improve short-term memory and, in turn, long-term memory. Rehearsing (or continually repeating) the information one wants to remember is a brute force approach to encoding the information into long-term memory. It causes information to be remembered in the practiced form and is effective given that other intellectual activities do not interfere with the rehearsals during the time required for the person to memorize the information.

"Mneumonics are conscious ways for helping to ensure the retention of material that would otherwise be forgotten . . . [they] are cognitive performance aids" (Bailey, 1989, p. 145). In general, mneumonics work by either reducing the amount of information that one needs to remember or by adding to it in a way that makes it easier to encode and retrieve. Acronyms represent a reducing method. For example, one might try to remember the five great lakes by using the one word acronym *homes* for Huron, Ontario, Michigan, Erie, and Superior. In contrast, the approach of using phrases, songs, concepts or visual imagery to remember information represents elaboration methods. For example, music teachers often use the phrase "Every good boy does fine," to help students remember the lines of the treble clef, E, G, B, D, and F.

Once encoded, stored information (i.e., long-term memory) represents an associative network of information and knowledge. To quote Hogarth (1987, pp. 134–135):

Although there is some controversy in the literature, most theorists now agree that at the moment of recall, long-term memory works largely by a process of reconstruction. In other words, the paradox of memory is that long-term memory does not work by remembering what is actually recalled, but rather by remembering fragments of information that allow one to construct more complete representations of the information. Moreover, these fragments of information can be thought of as being linked in a net-like collection of associations. The richer the associations triggered by specific information, the more likely people are to be able to recall that information.

Long-term memory is critical to judgment and decision processes in many ways. For example, it affects the manner in which the inference- and decision-making task is structured or framed. It also includes the type of cues and goals that a person selects as relevant in a particular situation. It affects the type of situations that are recognized. This implies a pattern-matching process whereby the current situation is considered similar to one that was previously experienced or learned or consistent with the person's schema for how events in a situation should go together. However, memory is also essential for function-relation processes, in which the general relations between cues and criteria represent the essential information in a situation.

In a similar vein, memory helps a person know what to do, that is, the type of decision to make and how to implement it. Memory is also critical in the interpretation and coding of outcomes or consequences of making certain decisions. It is also critical to guiding future expectations and action preferences. This process is more generally referred to as *learning*.

Memory limitations (in terms of encoding, storage, and retrieval) force people to use cognitive simplification mechanisms that are efficient in dealing with environmental complexity, but leave one open to possible cognitive biases. Some of these biases have been discussed, such as availability of instances in memory, selective perception of information, representativeness of information, and hindsight. Three other findings deserve mention here. First, people are more likely to remember concrete information (e.g., incidents) than abstract, statistical information (e.g., base rates). However, this finding appears to be mediated by the meaningfulness of the information. Second, people are more likely to remember information that has been coded on several sensory dimensions, for example, words and images versus only words. Third, when perceptions are incongruent with expectations, memory distortions can occur.

We close this brief review by discussing two findings regarding the retrieval of information from memory. First, recognizing information is much easier than recalling it from long-term memory. Therefore, interfaces should be designed to facilitate recognition, not recall. Second, memory may be inaccessible because the retrieval strategies employed in a particular context interfere with the strategies used during learning. There are two types of interference. Proactive interference is when the learning of a previous task interferes with use of newly learned material, that is, the old interferes with the new. Retroactive interference is when the learning of a new task interferes with the use of previously learned material, that is, new interferes with old. However, interference can be thought of as a general concept. There are a lot of ways that ongoing activities can interfere with one's ability to remember something. This interference creates a set of associations that is inconsistent with the set of associations that is needed to retrieve the relevant information.

Cognitive Science Findings: Attention

Research strongly supports the common-sense notion that the type of displays can significantly affect the information to which one attends. More generally, we can categorize three types of problems considered in the literature:

- Selective Attention: Person focuses too much on one thing and fails to attend to other critical information
- Focused Attention: The external environment forces a person to attend to certain information and not any other
- Divided Attention: The external environment forces a person, or the person tries, to process information for multiple tasks at the same time.

Workload conditions can, in various ways, cause these problems.

Regarding selective attention, we assume that a person is receiving information over multiple channels. We also assume that the person can selectively attend to these information channels. Now, what does the research say about the optimality of people's selection process?

The first conclusion is that the sampling of information channels is guided by an internal model of the statistical properties of the environment, for example, the probability that important information is coming over a channel or the mean time between pieces of information over a channel(s). Second, people learn to sample channels with higher event rates (i.e., frequency of important information) more frequently. However, sampling is not optimal. Third, there can be many reasons for why sampling is not optimal. For example, memory limitations can cause sampling problems; one simply forgets when a channel(s) was last sampled. Or something in the environment causes one to selectively attend to certain channels; this is *focused attention.* Or the person may not know the statistical properties of the environment. In this regard, research shows improvements in channel sampling when people are presented with a preview of scheduled events. This is a form of cognitive feedback.

Under high stress and arousal, which typically accompanies high workload conditions, the focus of attention restricts, that is, fewer channels and cues are sampled. "Fortunately and adaptively, the channels that are sampled generally tend to be those that are perceived by the subject as being most important. Importance here is determined by the frequency of events—their costs or values and their salience" (Wickens, 1984, p. 252). Salience represents a potential problem: "Salient cues (those in central vision or those that are intense) may not necessarily be the best ones to sample. An important objective of engineering psychology then should be to establish ways to ensure that all important channels and not just the salient ones will be sampled in times of high stress" (p. 253).

When considering focused versus divided attention, one should distinguish between serial and parallel processing. Specifically, focused attention (i.e., the external environment causes one to selectively attend to certain information channels and sequentially process information on that channel) is only a problem when one should be processing information in parallel (i.e., attending to two or more pieces of information coming in at the same time or place). Obviously this can occur under conditions of high workload. Conversely, divided attention (i.e., attempting to parallel process) is only a problem when one should be selectively attending to and serially processing certain information. Again high workload conditions, but now where the demands are coming from multiple, diverse sources, can result in attention problems.

It is difficult to process information in parallel when it comes from different locations in the visual field. Trying to make people do so often leads to selective attention and serial processing; that is, people focus on one location (or channel) in the visual field, process the information being transmitted there, and then move on to process the information being transmitted in the other location in the visual field.

If the operator must attend to a number of information channels, then parallel processing will enhance the probability of detecting an event on one of the channels. One way to enhance parallel processing is to have two sources of information come over one channel, that is, to merge two channels into one. This merging can be helpful (i.e., facilitate parallel processing) or harmful, depending on the consistency of the information sources and their implications for operator action.

It is helpful to present information about multiple dimensions over a single information channel when the information is highly correlated or presents an integrated view of an object. This can be accomplished by using both color and shape to represent information. It also can be accomplished by using object representations on the display. To quote Wickens (1984, p. 1969):

> When the purpose of a display is to provide information to an operator concerning the status of an object (e.g., the position and attitude of an aircraft, vehicle, or remote manipulator robotic arm), there is an intrinsic advantage to displaying the information concerning the object in an object format rather

than just in terms of its disembodied separate attributes—for example, as separate gauges or meters.

Similarly, when two information sources have independent implications for action, then parallel processing is helpful. "For example, the pilot controlling both pitch and roll of an aircraft (two relatively independent dimensions) will benefit from parallel processing when these are displayed close together" (p. 256). However, stimuli with incompatible implications for action or incompatible mental processes will cause interference if presented on the same channel. More generally, any presentation of information on the same channel that causes "resource competition" will negatively affect information processing. This leads one to the concept of time-sharing mental resources.

Instead of assuming that there is a single, common pool of mental resources for attending to and processing information, Multiple Resource Theory (MRT) argues that there are different mental resources and that these resources can be differentiated. In general, the greater the structural dissimilarity between two competing (or simultaneous) tasks, the less likely it is that they will compete for common resources and thereby interfere with one another. For example, "the resources used for perceptual and central processing activities appear to be the same and functionally separate from those underlying the selection and execution of responses" (Wickens, 1984, p. 302). Similarly, divided attention between the eye and the ear is quite good, although there is a general preference for processing information with the visual modality. Finally, manual and vocal responses can be time-shared because manual responses are usually spatial in nature and therefore involve the right hemisphere. In contrast, vocal responses are verbal and involve the left hemisphere. Resources underlying control by the left hand are also in the right hemisphere; right-hand control is in the left.

This completes our overview of the "cognitive bases of design." The aforementioned findings can be used as a filter through which designers can pass hypotheses about how information technology might enhance cognitive processing and performance.

Information Processing Technology for Cognitive Systems Engineering

There are a variety of tools, methods, techniques, devices, and architectures available to the UCI designer; many more will emerge as we move toward the 21st century. The challenge—as always—lies in the extent to which designers can match the right tool or method with the appropriate problem. This section looks at a number of technology options now available to the designer, options that will evolve quite dramatically during the next 5 to 10 years. More specifically, the section examines the information technologies available to the cognitive systems engineer.

MODELS AND METHODS

Figures 3.1, 3.2, and 3.3 from Sage and Rouse (1986) suggest how several of the major methods classes can be described and assessed. "Assessment," of course, is the key. Designers must know precisely what method (or methods) to apply to which decision support requirements.

Over the past few years the design community has seen the preeminence of knowledge-based tools and techniques, although the range of problems to which heuristic solutions apply is much narrower than first assumed. It is now generally recognized that artificial intelligence (AI) can provide knowledge-based support to well-bounded problems where deductive inference is required (Andriole, 1990). We now know that AI performs less impressively in situations with characteristics (expressed in software as stimuli) that are unpredictable. Unpredictable stimuli prevent designers from identifying sets

METHOD / CRITERIA	Artificial Intelligence	Cognitive Science	Decision Analysis	Operations Research and Control Theory
Objectives	Software Performance	Explanation Of Cognition	Aiding Decisions	Prediction and Optimization
Methods	Symbolic; Powerful Software Tools	Symbolic; Computation-Oriented	Numerical; Axiomatic Dec. Rules	Numerical; Predictive Models
Products	Software and Tools	Theories; Training and Aiding Princ.	Decision Prescriptions; Software/Tools	Optimal Solutions
Strength	Exploitation of Context; Software Tools	Emphasis on Process Over Product of Behavior	Well-Founded and Structured	Breadth and Rigor
Weaknesses	Often Ad Hoc; Avoids Engineering Methodology	Often Ad Hoc; Avoids Psychological Methodology	Biased Toward Choice; Not Judgment in Context	Avoids Context; Assumption Laden

FIG. 3.1. Methods assessment. From Sage & Rouse (1986).

of responses and therefore limit the applicability of if–then solutions. We now know, for example, that so-called expert systems can solve low-level diagnostic problems, but cannot predict Russian intentions toward Poland in 1998. Although there were many who felt from the outset that such problems were beyond the applied potential of AI, there were just as many who were sanguine about the possibility of complex inductive problem solving.

The latest methodology to attract attention is neural network-based models of inference making and problem solving. Neural networks are applicable to problems with characteristics that are quite different from those best suited

METHOD / CRITERIA	AI & Expert Systems	Decision Analysis & Decision Support	Human/System Interaction
Objectives/ Expectations	Design of "Intelligent" Systems (KBS) Human-like Capacities Heuristics	Normative Modeling, Aids to Option Generation, choice & selection	Effective Interface - (Adaptive, intelligent?) Between Humans and Decision Tools - Transparent Interface
Typical Analytical Concern or Question to Subject	"Models" of Human Processes, Search & Representation, Symbolic Processing	Decomposition, structure - elicitation of Subjective Judgements, Evaluation of Cost/Benefit MAU, Bayes Policy Capturing	Prototyping; Performance Modeling
Products	LISP Machines, Shells	Process Aids (Computer, Other)	I/O Devices, Configurations
Strengths: Input		Structure	
Process	Explicit Model, Ad Hoc	Normative Models, Face Validity of Models	
Output	Explanation Facility	Audit Trail Sensitivity Analysis	
Weaknesses: Input	Primitive	Often Data Intensive	
Process	Comprehensibility, May Use Unrealistic Belief Structures	Normative Models	
Output	Comprehensibility	Sensitivity Analysis	

FIG. 3.2. Methods assessment: options foci. From Sage & Rouse (1986).

METHOD / CRITERIA	OR & Control Engineering	Data Base Management	Cognitive Science/ Psychology
Objectives/ Expectations	Provide Results of Quantitative Analysis to Clients	Facilitate Storage, Retrieval, Manipulation of Data (Information)	Understanding & Describing Human Cognition
Typical Analytical Concern or Question to Subject	Linear Programs, Math Modeling, Optimization, Dynamic Systems Analysis, Statistics	Relational, Hierarchical, Spatial Structuring, Retrieval Strategies	Modeling Theories Lens Model Attribution Theory Empirical/Experimental
Products	System Model "Autopilot," etc. (Control Systems) Aids Modeling/Simulation Languages	DBMS Software, Database Designs	Human Models/Decision Heuristics (Validated?) Capability and/or Limitation Assessments
Strengths: Input	Can Use Massive Real Data		
Process	Rigorous; Engineering Methods		Experimental/Empirical
Output	Quantification		
Weaknesses: Input	Requires Volumes of Data	Inflexible	
Process	"Artificial" Lack of Face Validity (GIGO)		Mind Reading, Laboratory Bound
Output	Explanation Facility		Narrow Domains

FIG. 3.3. Methods assessment: options foci (*cont'd*). From Sage & Rouse (1986).

to AI. *Neural networks* are—according to Hecht-Nielsen (1988, p. 32)—"computing systems made up of a number of simple, highly interconnected processing elements which process information by their dynamic state response to external inputs." Neural nets are nonsequential, nondeterministic processing systems with no separate memory arrays. Neural networks, as stated by Hecht-Nielsen, comprise many simple processors that take a weighted sum of all inputs. Neural nets do not execute a series of instructions, but rather respond to sensed inputs. Knowledge is stored in connections of processing elements and in the importance (or weight) of each input to the processing elements. Neural networks are allegedly nondeterministic, non-algorithmic, adaptive, self-organizing, naturally parallel, and naturally fault tolerant. They are expected to be powerful additions to the methodology arsenal, especially for data-rich, computationally intensive problems. The intelligence in conventional expert systems is preprogrammed from human expertise, whereas neural networks receive their "intelligence" via training. Expert systems can respond to finite sets of event stimuli (with finite sets of responses), whereas neural networks are expected to adapt to infinite sets of stimuli (with infinite sets of responses). It is alleged that conventional expert systems can never learn, whereas neural networks "learn" via processing. Proponents of neural network research and development have identified the kinds of problems to which their technology is best suited: computationally intensive, nondeterministic, nonlinear, abductive, intuitive, real-time, unstructured/imprecise, and nonnumeric (DARPA/MIT, 1988).

It remains to be seen if neural networks constitute the problem-solving panacea that many believe they represent. The jury is still out on many aspects of the technology. But like AI, it is likely that neural nets will make a measured contribution to our inventory of models and methods.

What does the future hold? Where will the methodological leverage lie? In spite of the overselling of AI, the field still holds great promise for the design and development of interactive, embedded, and real-time systems. Natural language processing systems—systems that permit free-form English interaction—will enhance decision support efficiency and contribute to the wide distribution of all kinds of systems. The Artificial Intelligence Corporation's INTELLECT natural language processing system, for example, permits users to interact freely with a variety of database management systems. Cognitive Systems, Inc.'s BROKER system permits much the same kind of interaction with the Dow Jones's databases. These systems are suggestive of how natural language interfaces will evolve over time, of how users will be able to communicate with databases and knowledge bases in ways that are compatible with the way they address human and paper data, information, and knowledge bases. When users are able to type or ask questions of their systems in much the same way they converse with human colleagues, then the way computer-based systems will be used will be changed forever.

Expert systems will also routinize many decision-making processes. Rules about investment, management, resource allocation, and office administration will be embedded in expert systems. Problems that now have to be resolved whenever a slight variation appears will be autonomously solved.

Smart database managers will develop necessary databases long before decision support problems are identified. Systems of the 1990s will be capable of adapting from their interaction with specific users. They will be able to anticipate problem-solving "style" and the problem-solving process most preferred by the user. They will be adaptive in real time and capable of responding to changes in the environment, such as a shortage of time.

The kinds of problems that will benefit the most from AI will be well-bounded, deductive inference problems about which a great deal of accessible and articulate problem-solving expertise exists. The community will abandon its goals of endowing computer programs with true inductive or abductive capabilities in the 1990s, and the dollars saved will be plowed back into so-called "low-level" AI.

The application of knowledge-based processing to the AWACS weapons direction process, for example, has been limited by design (see Chapter 4 for more details). First and foremost, the application—quasi-automated nomination of assets to targets—falls within a well-bounded domain. There are only a finite number of pairing options that any human operator or computer-based system could consider given a set of definable situational characteristics. We thus implemented a low-level knowledge-based solution to one of the major weapons direction requirements.

Designers in the 1990s will also benefit from a growing understanding of how humans make inferences and decisions. As the Phase I and Phase II projects have demonstrated, the cognitive sciences are amassing evidence about perception, biasing, option generation, and a variety of additional phenomena directly related to decision support systems (DSS) modeling and problem solving. The world of technology will be informed by new findings; resultant systems will be "cognitively compatible" with their users.

Next generation systems will also respond to the situational and psychophysiological environment. They will alter their behavior if their user is making a lot of mistakes, taking too long to respond to queries, and the like. They will slow down or accelerate the pace, depending on this input and behavior. The field of cognitive engineering—which will inform situational and psychophysiological system design strategies—will become increasingly credible as we approach the 21st century. The traditional engineering developmental paradigm will give way to a broader perspective that will define the decision-making process more from the vantage point of requirements and users than computer chips and algorithms. Principles of cognitive engineering will also inform the design and human computer interfaces (see later).

Some future support software will be generic, and some will be problem specific. Vendors will design and market generic accounting, inventory control, and option selection software. These models will be converted into templates that can be inserted directly into general-purpose systems. The template market will grow dramatically over the next 5 to 10 years.

It is extremely important to note the appearance of development tools. Already there are packages that permit the development of rule-based expert systems. There are now fourth generation tools that are surprisingly powerful and affordable. These so-called "end-user" systems will permit on-site design and development of DSSs that may only be used for a while by a few people. As the cost of developing such systems fails, more and more throwaway systems will be developed. This will change the way we now view the role of decision support in any organization, not unlike the way the notion of rapid application prototyping has changed the way application programs should be developed.

Hybrid models and methods drawn from many disciplines and fields will emerge as preferable to single model-based solutions, largely because developers will finally accept diverse requirements specifications. Methods and tools drawn from the social, behavioral, mathematical, managerial, engineering, and computer sciences will be combined into solutions driven by requirements and not by methodological preferences or biases. This prediction is based in large part on the maturation of the larger design process, which today is far too vulnerable to methodological fads. Hybrid modeling for DSS design and development also presumes the rise of multidisciplinary education and training, which is only now beginning to receive serious attention in academia and industry.

USER-COMPUTER INTERFACE (UCI) TECHNOLOGY

Twenty years ago no one paid much attention to user interface technology. This is understandable given the history of computing, but no longer excusable. Since the revolution in microcomputing—and the emerging one in workstation-based computing—software designers have had to devote more attention to the process by which data, information, and knowledge are exchanged between the system and its operator. There are now millions of users who have absolutely no sense of how a computer actually works, but they rely on its capabilities for their very professional survival. A community of "third party" software vendors is sensitive to both the size of this market and its relatively new need for unambiguous, self-paced, flexible computing. It is safe to trace the evolution of well-designed human–computer interfaces to some early work in places such as the University of Illinois, the Massachusetts Institute of Technology (in what was then the Architecture Machine Group, now the Media Lab), Xerox's Palo Alto Research Center (Xerox/Parc), and, of course, Apple Computer, Inc. The "desk-top" metaphor, icon-based navigational aids, direct manipulation interfaces, and user-guided or controlled interactive graphics—among other innovations—can all be traced to these and other organizations.

Where did all these ideas come from? The field of cognitive science and now cognitive engineering is now—justifiably—taking credit for the progress in UCI technology, because its proponents were the (only) ones asking why the user–computer interaction process could not be modeled after some validated cognitive information-processing processes. UCI models were built and tested, and concepts such as spatial database management (from MIT's Architecture Machine Group; Bolt, 1984), hierarchical data storage, and hypertext were developed. It is no accident that much UCI progress can be traced to findings in behavioral psychology and cognitive science; it is indeed amazing that the cross-fertilization took so long.

UCI progress has had a profound impact on the design, development, and use of interactive systems. Because many of the newer tools and techniques are now affordable (because computing costs have dramatically declined generally), it is now possible to satisfy complex UCI requirements even on personal computer-based systems. Early data-oriented systems displayed rows and rows (and columns and columns) of numbers to users; modern systems now project graphic relationships among data in high resolution color. Designers are now capable of satisfying many more substantive and interface requirements because of what we have learned about cognitive information processing and the affordability of modern computing technology.

The most recent progress in UCI technology is multimedia, or the ability to store, display, manipulate, and integrate sound, graphics, video, and good old-fashioned alphanumeric data (Aiken, 1989; Ambron & Hooper, 1987; Ragland, 1989). It is now possible to display to users photographic, textual, numerical, and video data on the same screen. It is possible to permit users to select (and deselect) different displays of the same data. It is possible to animate and simulate in real time—and cost-effectively. Many of these capabilities were just too expensive a decade ago and much too computationally intensive for the hardware architectures of the 1970s and early 1980s. Progress has been made in the design and execution of applications software and in the use of storage devices (such as videodisks and compact disks [CDs]). Apple Computer's Hypercard software actually provides drivers for CD players through a common UCI (the now famous "stack"). Designers can exploit this progress to fabric systems that are consistent with the way their users think about problems. There is no question that multimedia technology will affect the way future systems are designed and used. The gap between the way humans "see" and structure problems will be narrowed considerably via the application of multimedia technology.

Direct manipulation interfaces (DMIs) such as trackballs, mice, and touch-screens have also matured in recent years and show every likelihood of playing important roles in the next generation UCI design and development. Although there is some growing evidence that the use of the mouse can actually degrade human performance in certain situations, there are countless others in which the payoff is empirically clear (Bice & Lewis, 1989; Ledgard, Singer, & Whiteside, 1981; Ramsey & Atwood, 1979). Touch screens are growing in popularity when keyboard entry is inappropriate and for rapid template-based problem solving (Smith & Mosier, 1984).

The use of graphical displays of all kinds will dominate future UCI applications. Growing evidence in visual cognition research (Pinker, 1985) suggests how powerful the visual mind is. It is interesting that many problem solvers—professionals who might otherwise use a computer-based support system—are trained graphically not alphanumerically. Military planners receive map-based training; corporate strategists use graphical trend data to extrapolate and devise graphic scenarios; and a variety of educators have taken to using case studies laden with pictures, icons, and graphics of all kinds. Complicated concepts are often easily communicated graphically, and it is possible to convert complex problems from alphanumeric to graphic form. There is no question that future systems will exploit hypermedia, multimedia, and interactive graphics of all kinds.

Speech input and output should also emerge over the next 5 to 10 years as a viable UCI technology. Although predictions about the arrival of voice-activated text processors have been optimistic to date, progress toward even continuous speech input and output should be steady. Once the technology is perfected, there are a number of special-purpose applications that will benefit greatly from keyboard- and mouseless interaction.

The use of advanced UCI technology will foster a wider distribution of interactive computer-based systems. Early systems were used most productively by those familiar with the method or model driving the system as well as interactive computing itself. In other words, in order to exploit technology one had to have considerable computing expertise. Advanced UCI technology reduces the level of necessary computing expertise. Evidence suggests that training costs on the Apple Macintosh, for example, are lower because of the common user interface. Pull-down and pop-up menus, windows, icons, and direct manipulation via a mouse or trackball are all standard interface equipment, regardless of the application program (and vendor). If you know how to use one Macintosh program, chances are you can use them all to some extent. Such interface uniformity is unheard of in other than Macintosh-based software systems, yet it illustrates the enormous leverage that lies with the creative application of advanced UCI technology.

UCI technology will also permit the use of more methods and models, especially those driven by complex—yet often inexplicable—analytical procedures. For example, the concept of optimization as manifest in a simplex program is difficult to communicate to the typical user. Advanced UCI technology can be used to illustrate the optimization calculus graphically and to permit users to understand the relationships among variables in an optimization equation. Similarly, probabilistic forecasting methods and models anchored in Bayes' Theorem of conditional probabilities, although computationally quite simple, are conceptually convoluted to the average user. Log odds and other graphic charts can be used to illustrate how new evidence impacts prior probabilities. In fact, a creative cognitive engineer might use any number of impact metaphors (like thermometers and graphical weights) to present the impact of new evidence on the likelihood of events.

Finally, advanced UCI technology will also permit the range of decision support to expand. Anytime communications bandwidth between system and user is increased, the range of applied opportunities grows. UCI technology permits designers to attempt more complex system designs due to the natural transparency of complexity that good UCI design fosters.

Some argue that the interface may actually become "the system" for especially inexperienced users. The innards of the system—like the innards of the internal combustion engine—will become irrelevant to the operator. The UCI will orchestrate process, organize system contents and capabilities, and otherwise shield users from unfriendly interaction with complex data, knowledge, and algorithmic structures.

HARDWARE

The hardware that supports information-processing technology today is conventional. There are turnkey systems as well as generic hardware configurations that support the use of alternative systems. CPUs, disk drives, keyboards, light pens, touch screens, and the like can be found in a variety of systems. There are also microcomputer systems, as well as systems that require larger (minicomputer) hardware configurations.

Next generation systems will be smaller and cheaper and therefore more widely distributed. They will be networked and capable of uploading and downloading to larger and smaller systems. Input devices will vary from application to application as well as from the preferences of the user. As suggested earlier, voice input will dramatically change the way a small set of systems are used in the future; voice-activated text processing will expand the capabilities of systems by linking decision support to word processing and report preparation in a "natural" unobtrusive way, although it is likely that robust voice-activated systems will not appear until the late 1990s.

Many systems will have embedded communications links to databases and knowledge bases, other systems on decision support networks, and the outside world via conventional communication systems. Future systems will have (some selected) voice-input capabilities, conventional headset communications, deep database linkages, and a "place" on a much larger decision support network.

Briefcase- and smaller sized systems will become widespread. The embedding of spreadsheets in popular portable microcomputers suggests that decision support chips will be developed and embedded in future hardware configurations. In fact, not unlike some of the more powerful calculators of the 1970s, future systems will permit users to mix and match support chips within a single processor. The chips will be sold independently by decision support vendors.

Future interactive, embedded, and real-time systems will also be integrated with video display systems of several genres. There will be video disk-based systems as well as packaged systems that integrate powerful computer-generated imagery capabilities. The cost of both video options is falling rapidly, and the consumer of the future will be able to select the one that best serves his or her needs.

It is safe to say that video will become integral to future decision support. But it will be integrated directly into the system, not as an optional, expensive afterthought.

Behavioral scientists have just about convinced system architects—via the amassing of tons of evidence—that information, concepts, and many ideas can be communicated much more effectively via graphic, symbolic, and iconic displays (Shneiderman, 1987; Smith & Mosier, 1984).

The revolution in high-resolution display technology will exert a profound impact on next generation systems design and use. Many UCI technologies will exploit high-resolution displays, thereby accelerating the movement toward graphic computing.

Processor technology is also evolving rapidly. Just a decade ago, most of us computed on Intel 8088 microprocessors, whereas today everyone is processing on 486s, Pentiums, and Power PCs. Processors like the Motorola 68030 and 68040 have placed enormous power not only in the hands of users, but—perhaps more importantly—with designers as well. It is to say that applications software today is lagging the capabilities of such chips; at the same time, even assuming a consistent lag, systems in the 1990s and beyond will benefit from applications software that exploits the revolution in microprocessor design.

The issue of power, however, does beg the larger requirements question. In other words, it is safe to assume that raw computing power will be ready for next generation concepts. The challenge—as always—will lie in the application of the power to validated user requirements. If the truth be told, there are many successful systems that today require less than 20% of available computational power; many future systems may well find themselves with abundant power—and nowhere to go! Regardless of available computing power, designers will have to adhere to sound information systems engineering principles well into the 1990s and into the foreseeable future (Andriole, 1990).

We are witnessing the demise of the distinction among mainframe, mini-, and microcomputers. Tomorrow there will be workstations. Some will be more powerful than others, but nearly all will be available to individuals at reasonable prices. The balance between capability and price will continue to perplex vendors, because users will demand more and more capabilities for less and less money. Pricing strategies will determine how much power becomes affordable. Future design and use will work within this changing marketplace and, because of some new usage strategies (see next section), will remain largely unaffected by the instability of the workstation marketplace.

USAGE PATTERNS

Systems will be used very differently in the future than they are today. They may well function as clearinghouses for our professional problems. They may prioritize problems for us, and they may automatically go ahead and solve some of them. They will become problem-solving partners, helping us in much the same way colleagues do now. The notions of intelligent systems as software or hardware and users as operators will give way to a cooperative sense of function that will direct the design, development, and application of the best systems.

They will also be deployed at all levels in the organization. Today, decision support is targeted at midlevel management; tomorrow all levels will be supported by powerful interactive, adaptive systems. The distribution of systems will permit networking, the sharing of decision support data, and the propagation of decision support problem-solving experience (through the development of a computer-based institutional memory of useful decision support "cases" that might be called on to help structure especially recalcitrant decision problems). Efficient organizations will actually develop an inventory of problem/solution combinations that will be plugged into their decision support networks.

Future systems will also communicate with systems in other organizations in other parts of the world. Falling satellite communications costs will permit global linkages and contact with databases, expert systems, inventories, and the like, thereby multiplying the capabilities of in-house systems by orders of magnitude. This global networking is not decades away, but only 5 to 10 years away.

The most important change will occur in the way systems interface with other information systems. Most contemporary systems are "disembodied," that is, distinct from larger corporate, government, or military information systems. Actual use of many systems involves leaving one system to activate another. It is common in the military application of decision support for users to work alternately with mini- and microcomputers, manually feeding the output from one system into the other. A good deal of this can be explained by acquisition and procurement craziness, but just as much can be traced to obsolete concepts of how systems should be used. As the range of problems and capabilities increases, fewer and fewer systems will be disembodied; to the contrary, the most successful systems will be embedded in larger organizational and executive information systems.

Future executive information systems will provide portals for users to explore. It will be possible to perform all sorts of tasks via myriad application programs (that ideally will have common user–computer interfaces).

Finally, next generation systems will bridge the gap between our professional and personal worlds. Because they will have capabilities to manage our professional lives, they will also be capable of managing our personal lives as well. Because future decision support "delivery systems" will be expandable and inexpensive, the integration of personal management modules will be inevitable. This blurring of the traditional lines between our professional and personal worlds may not be desirable, but it is likely. Again, the technology will drive such changes to a significant extent, technology that will trigger changes because of its capabilities not necessarily in response to real requirements. In other words, there is a technological destiny built into our forecasts about the role of future interactive, embedded, and real-time systems.

The net effect is staggering. Decision support in the 1990s will be enormously broad and powerful. It will be distributed and networked. It will be intelligent and inexpensive. The effects of this reality are difficult to precisely predict, although a number of the ideas expressed here are in fact inevitable. Have we given enough thought to the direction in which decision support technology is taking us? Have we assessed the desirability of the direction? Have we determined the impact that next generation systems will have on the office of the future? Will they define the office of the future? Or will the office of the future suggest a role for decision support? We have avoided these and other tough questions. If we are to manage and exploit the new technology, we need to address all of the pertinent questions as soon as possible. When we do, we will be able to inform and direct progress so that the next generation is as capable and responsive to real needs as it can possibly be.

The technology tour presented here suggests that designers will leverage emerging technology to fundamentally change the nature of interactive decision support. Although the recent past enjoyed data-oriented decision support, next generation systems will provide analytical support of all kinds. Of particular value will be the speed with which routine problems will be solved and the advisory support that systems will provide users in areas as complex as option generation, evaluation, and selection. Interface technology will play a pivotal role in the introduction of new and hybrid models and methods; the hardware community will make sure that more than enough computational power is available.

The whole concept of decision support will evolve to accommodate changes in the larger corporate, governmental, and military information systems structure. Networking and advanced communications technology will permit linkages to databases and knowledge bases—and the routines to exercise them. Not only will distinctions among mainframe, mini-, and microcomputing fade, but distinctions among management information, executive information, and decision support systems will also cloud. Ironically, the concept of centralization may reappear, not with reference to central computing facilities, but with regard to enormous systems conceived functionally as hierarchies of capabilities. Users may well find themselves within huge computing spaces capable of supporting all kinds of problem solving. Advanced communications technology will make all this possible; users will be able to travel within what will feel like the world's largest mainframe, which conceptually is precisely what a global network of data, knowledge, and algorithms is.

The same users will be able to disengage the network and go offline to solve specific problems. This freedom will expand the realm of analytical computing in much the same way microcomputing expanded the general user community.

Finally, all this technology will permit designers to fulfill user requirements in some new and creative ways. Up until quite recently, technology was incapable of satisfying a variety of user requirements, simply because it was too immature or too expensive. We have crossed the capability/cost threshold; now designers can dig into a growing tool bag for just the right methods, models, and interfaces. By the year 2000, this tool bag will have grown considerably. Talented designers should be able to match the right tools with the right requirements to produce systems that are user oriented and cost effective.

The future of systems design, development, and use is bright. Although some major changes in technology and application concepts are in the wind, next generation systems will provide enormous analytical support to their users. We can expect the range of decision support to grow in concert with advances in information technology.

Case Studies
in Context

This book argues that empirical findings from cognitive science can be leveraged in the design and development of user–computer interfaces likely to enhance human performance.

In the larger field of human–computer interaction or user interface design, cognitive systems engineering is part of the larger ongoing quest to make the interface between organic and electronic systems smoother, friendlier, and more productive.

Table 4.1 presents Chignell's user interface taxonomy (Chignell, 1990). This taxonomy locates cognitive engineering in the larger endeavor and suggests (a) the importance of the area, and (b) where this book—and its case studies—are focused.

CASE STUDIES

The Air Defense Intelligence and Operations
"Conceptual Case Study"

The case studies themselves crosscut a number of domains and tasks. It is important to note that we began this research with a conceptual case study. We analyzed the potential impact of cognitive systems engineering on an air defense and intelligence domain and developed an interactive prototype interface based directly on findings from cognitive science. The domain called for supporting requirements to monitor and understand situations, make inferences, and generate options.

TABLE 4.1
Chignell's UCI Taxonomy

The Basic Interface Model
The first branch reflects the view that the user interface is composed of the seven fundamental components:
1.1 Actions
1.2 Behaviours
1.3 Contexts
1.4 Displays
1.5 Effects
1.6 Forms
1.7 Goals
Each of these fundamental components may then be further expanded into subcomponents. The actions of the computer application may be broken down into traditional components of processing, i.e., cpu, I/O, and peripherals, leading to the following categorization:
1.1 Actions
    1.1.1 Computation
    1.1.2 Storage
    1.1.3 Retrieval
    1.1.4 Operation of Peripheral Devices
The behaviors of the user are a little bit more difficult to characterize. In this classification, we have chosen to emphasize tradition concerns in HCI, i.e., the problems of navigation, information seeking, interaction styles, and input devices. Of these four problems, the issue of user navigation is probably the least understood. We will provisionally characterize user navigation as the way in which the user moves around the conceptual structure of the interface. In practice, there will be a high degree of overlap between user navigation and interaction styles, since the intentions of the user during navigation will have to be communicated to the interface via some style of interaction.
1.2 Behaviours
    1.2.1 User Navigation
        1.2.1.1 Selection from Map or Display
        1.2.1.2 Move to more general command or concept
        1.2.1.3 Move to more detailed command or concept
        1.2.1.4 Move to associated concept or command
        1.2.1.5 Switch between modes, windows or environments
        1.2.1.6 Ask for Help
        1.2.1.7 Move through a sequence of commands or information
    1.2.2 Interaction Styles
        1.2.2.1 Direct Manipulation
            1.2.2.1.1 Icons
            1.2.2.1.2 Engagement
            1.2.2.1.3 S-R Compatibility
        1.2.2.2 Command Driven
            1.2.2.2.1 Command Languages
                1.2.2.2.1.1 Command Syntax
                1.2.2.2.1.2 Language Semantics and Expressiveness
                1.2.2.2.1.3 Shortcuts and Macros
            1.2.2.2.2 Command Menus
        1.2.2.3 Form Filling
            1.2.2.3.1 Query by Example

(Continued)

TABLE 4.1
(Continued)

1.2.2.4 Menus
    1.2.2.4.1 Menu Design
    1.2.2.4.2 Menu Maps
    1.2.2.4.3 Menu Selection
1.2.2.5 Conversational
1.2.2.6 Graphical Structures
1.2.3 Input Devices
    1.2.3.1 Data Entry
        1.2.3.1.1 Keyboard
        1.2.3.1.2 Function Keys
        1.2.3.1.3 Bar Codes
        1.2.3.1.4 Digitizing Tablet
    1.2.3.2 Pointing Devices
        1.2.3.2.1 Direct Pointing
            1.2.3.2.1.1 Lightpen
            1.2.3.2.1.2 Touch Screen
        1.2.3.2.2 Indirect Pointing
            1.2.3.2.2.1 Mouse
            1.2.3.2.2.2 Trackball
            1.2.3.2.2.3 Joystick
            1.2.3.2.2.4 Graphics Tablet
1.2.3.3 Speech Recognition

The next category (contexts) is an aspect of the basic interface model that has not received much attention in HCI research. What are the types of contexts that user behaviors can occur in? It turns out that there are many different ways to classify the kinds of contexts that occur in HCI. We will list a few that seem to have a strong effect on performance. These include; task complexity, temporal constraints, hardware or software malfunctions, knowledge of results (i.e., are the task relevant effects of previous behavior visible to the user?)

1.3 Contexts
    1.3.1 Task Complexity
    1.3.2 Temporal Constraints
    1.3.3 System Malfunctions, Limitations, and Capabilities
    1.3.4 Knowledge of Results

The category of displays deals with a new and relatively unknown area. While the general problem of displays has been around for a long time, recent technological advances have made it possible to combine information from a number of different media and to compose it so that the screen and sound spectrum can become a collage of information from different sources. Provisionally, we have divided the category of displays into the topics of multimedia, screen design, and display composition, recognizing that there will be some overlapping issues between these subcategories.

1.4 Displays
    1.4.1 Multimedia
        1.4.1.1 Auditory Displays
        1.4.1.2 Video
        1.4.1.3 2-D Graphics
        1.4.2.4 3-D Graphics
        1.4.2.5 Text Display

(Continued)

TABLE 4.1
(Continued)

The distinction between displays and effects reflects the difference between information presented for its own sake and feedback about the operations of the application in response to user behaviors. Many of the details about effects will be application specific, so our classification will list only those types of effect that generally occur across applications.

As stated earlier, forms are the models in which actions, effects, and displays are embedded. The role of visual metaphors and spatial models as types of form seems obvious. However, visualizing information has also been included as a form because it deals with the task of making the user understand the structure and gross features of information rather than the details of the information. This is consistent with the general role of forms which is to assist the user in forming an accurate conceptual model of how the interface, and the information and functions that it contains, is structured. Examples of the work on visualizing information include use of automatic icons (Fairchild, Meredith, and Wexelblat, 1988a) and artificial realities (Fairchild, Meredith, and Wexelblat, 1988b).

(Continued)

TABLE 4.1
(Continued)

1.6.3 Visualizing Information
    1.6.3.1 Automatic Icons
    1.6.3.2 Simulations and Artificial Realities

Goals represent the motivating forces behind HCI. Task analysis is the process by which a task is decomposed into sequences of goals and subgoals. It is typically a normative process based on a rational model of the task. In contrast, goal identification is a descriptive process that infers goals on the basis of user behavior while performing the task. Methods for goal identification include verbal protocol analysis, interviewing, error and critical incident analysis, and response time studies, where goals are inferred on the basis of preparatory pauses in transaction logs (e.g., Eberts, 1987).

1.7 Goals
    1.7.1 Normative Task Analysis
    1.7.2 Descriptive Goal Identification
        1.7.2.1 Verbal Protocol Analysis
        1.7.2.2 Interviewing
        1.7.2.3 Error and Critical Incident Analysis
        1.7.2.4 Response Time Studies

Cognitive Engineering

The second branch of the tree is cognitive engineering. This refers to the process of applying the models and findings of cognitive science to the task of analyzing and designing user interfaces. The cognitive engineering branch is broken down into three sub-branches:

2.1 Cognitive Science
2.2 Normative Models
2.3 Descriptive Models

The three sub-branches of cognitive engineering may then be further broken down into subtopics as shown below:

2.1 Cognitive Science
    2.1.1 Cognitive and Experimental Psychology
        2.1.1.1 Experimental Design and Analysis
        2.1.1.2 Human Intelligence and Abilities
        2.1.1.3 Personality and Motivation
        2.1.1.4 Human Information Processing
            2.1.1.4.1 Learning
            2.1.1.4.2 Memory
            2.1.1.4.3 Decision Making
            2.1.1.4.4 Problem Solving
            2.1.1.4.5 Attention
                 2.1.1.4.5.1 Selective Attention
                2.1.1.4.5.2 Divided Attention
                2.1.1.4.5.3 Focused Attention
                2.1.1.4.5.4 Attentional Resources
        2.1.1.5 Perception
    2.1.2 Artificial Intelligence
        2.1.2.1 Knowledge Representation
            2.1.2.1.1 Semantic Nets
            2.1.2.1.2 Frames
            2.1.2.1.3 Production Rules
            2.1.2.1.4 Scripts

(Continued)

TABLE 4.1
(Continued)

2.1.2.2 Symbolic Programming
      2.1.2.2.1 Lisp
      2.1.2.2.2 Prolog
      2.1.2.2.3 SmallTalk
2.1.2.3 Knowledge Engineering
      2.1.2.3.1 Knowledge Acquisition
      2.1.2.3.2 Machine Learning
      2.1.2.3.3 Inference
2.1.2.4 Machine Vision
2.1.3 Language Understanding
      2.1.3.1 Syntax
      2.1.3.2 Semantics
      2.1.3.3 Discourse Analysis
      2.1.3.4 Text Analysis
      2.1.3.5 Language Acquisition
2.1.4 Neuroscience
      2.1.4.1 Neurophysiology
      2.1.4.2 Action
      2.1.4.3 Human Vision
      2.1.4.4 Learning and Memory
      2.1.4.5 Hemispheric Differences
2.1.5 Philosophy
      2.1.5.1 Ontology
      2.1.5.2 Epistemology

At present, the section of normative models simply lists some of the more prominent approaches. Eventually, it would be desirable to classify these approaches into different types of model, e.g., high-level vs. low-level models.

2.2 Normative Models
      2.2.1 GOMS
      2.2.2 Model Human Processor
      2.2.3 Keystroke-level Model
      2.2.4 ACT*
      2.2.5 Task Action Grammar

Descriptive models of HCI have received a lot of attention in recent years. Basic distinctions can be made between the user's mental model of the system, the conceptual model of the system as seen by the designer, and the user model that the system has in interpreting user inputs and handling the interaction. Research methods for identifying these descriptive models tend to be the same as for those used in goal identification (1.7.2 above).

The subcategory of knowledge in this section of the taxonomy refers to methods for handling descriptive knowledge that is relevant in an HCI application. Knowledge acquisition and knowledge engineering are terms that have been used with reference to building expert systems, but they can also be applied to processes for discovering and using knowledge within HCI. Knowledge compilation is the process of streamlining and automating knowledge for particular tasks (Anderson, 1983). In terms of HCI, it can be used to describe both the compilation of the users knowledge about the application and the interface and the system's knowledge about the user and the task.

2.3 Descriptive Models
      2.3.1 Mental Model

(Continued)

TABLE 4.1
(Continued)

2.3.2 Conceptual Model
2.3.3 User Model
    2.3.3.1 Usage Profiles
    2.3.3.2 Resources and Workload
    2.3.3.3 Expertise
2.3.4 Domain Model
    2.3.4.1 Knowledge Acquisition
    2.3.4.2 Knowledge Engineering
    2.3.4.3 Knowledge Compilation

Interface Engineering

The next branch of the tree to be dealt with is interface engineering. This is the portion of the user interface research taxonomy which deals with guidelines and approaches for interface design (engineering). The five branches of interface engineering are:

3.1 General Guidelines
3.2 Interface Structuring
3.3 Interface Training
3.4 Interface Evaluation
3.5 User Interface Development Systems and Tools
3.6 Interface Engineering Techniques

The general guidelines represent a conceptual framework that guides HCI research and design. Some of these guidelines are based on empirical findings, others represent a consensus view of the HCI research community. Some of the guidelines have even reached the stage of being axiomatic. For instance, structuring information into manageable chunks (e.g., menu selections) and hiding unnecessary information from the user are close to being axioms of HCI.

The general guidelines shown below should be recognizable to most HCI researchers. The Do's and Don'ts refer to various lists of guidelines that have been published (e.g., Smith and Mosier, 1986). Computer anthropomorphism refers to the problem of making the computer seem more like a person than it really is, which can eventually lead to failed expectations and frustration on the part of the user.

3.1 General Guidelines
    3.1.1 Do's and Don'ts
    3.1.2 Excess Functionality
    3.1.3 Cognitive Compatibility
    3.1.4 Computer Anthropomorphism
    3.1.5 User-Centered Design

Interface structuring is a category that is somewhat related to the issue of user navigation (1.2.1 above). Possibilities for user navigation generally depend on the structuring principles and types of conceptual structuring that are available in the interface. Dialogue design is an aspect of conceptual structuring concerned with how users make choices, enter data and the like. The results of a dialogue will often be a transition from one state or situation to another.

The dominant structuring principles at present in HCI are hierarchies (e.g., menu systems), linear sequences, and networks or hypermedia. Look up or relational tables may also be used for structuring. Under the others subcategory might be included more specialized structuring methods such as spreadsheets.

3.2 Interface Structuring
    3.2.1 Conceptual Structuring
        3.2.1.1 Dialogue Design
        3.2.1.2 States or Situations

(Continued)

TABLE 4.1

(Continued)

3.2.2 Structuring Principles
    3.2.2.1 Hierarchies
    3.2.2.2 Hypermedia
    3.2.2.3 Linear Sequences
    3.2.2.4 Tables
    3.2.2.5 Others

The usability of an interface will be determined by the combination of its design, the task complexity, the user skills, and the availability of interface training, among other things. Interface training is a broad category, but relevant aspects include online tutorials, help facilities, and documentation.

3.3 Interface Training
    3.3.1 Online Tutorials
    3.3.2 Help Facilities
    3.3.3 Documentation

Interface evaluation is one of the most vexing issues in HCI. In general one can distinguish between formative evaluation, where one is seeking to improve the design, and summative evaluation, where one is trying to assign an overall figure of merit to a developed application. Face Validity refers to whether or not the proposed interface design has an acceptable look and feel (particularly to potential users). An adequate first impression of the proposed interface can often be conveyed with storyboards and similar techniques. Wizard of Oz experiments are used to construct a simulation of how the interface will behave prior to detailed programming. An example of a Wizard of Oz experiment would be a situation where the application is simulated by having an experimenter type in input to the user's terminal in accordance with the design. Prototypes may also be used to evaluate proposed designs. Rapid construction of prototypes is often possible using the methods listed in section 3.5 below.

For summative analysis of the interface once it has been constructed, the preferred method is performance analysis. Yet it may sometimes be difficult to specify an adequate benchmark task that provides an adequate test of the usability of the interface. It helps if there are alternative implementations of the application to compare the current interface with, since it is otherwise difficult to separate the effects of the interface from the effects of task complexity and the design of the application.

As a result of difficulties in performance analysis, questionnaires are probably the most frequently used method of summative evaluation of user interfaces. However, questionnaires provide a subjective evaluation of interfaces which is often greatly influenced by the type of questions asked and the way in which the questions are phrased.

One issue in summative evaluation is "when to evaluate?" An interface that the user initially finds difficult to use may eventually prove to be easy to use and have high functionality for the user. Learning curve analysis looks at how use of the interface improves over time. Thus instead of analyzing performance or administering questionnaires at a single point in time, several evaluations may be made over a period of time as the user becomes more familiar with the system. Thus learning curve analysis takes the other methods of summative evaluation and applies them over a time period to get a more complete profile on how usability changes with usage. In most situations the usability of an interface on the first day of use will not be as important as the slope of the learning curve over subsequent days and the amount of proficiency attained by users after extended use.

3.4 Interface Evaluation
    3.4.1 Formative Evaluation
        3.4.1.1 Face Validity

(Continued)

TABLE 4.1

(Continued)

3.4.1.2 Wizard of Oz experiments

3.4.1.3 Prototype Evaluation

3.4.2 Summative Evaluation

    3.4.2.1 Performance Analysis

    3.4.2.2 Learning Curve Analysis

    3.4.2.3 Questionnaires and Interviews

In the past, user interfaces have been notoriously difficult to construct. Software developers have often found that the majority of time spent in implementing an application is actually spent on the interface. Thus there has been a great deal of interest in the development of systems and tools that can both speed up the development of user interfaces and increase their level of usability. The taxonomy currently divides these systems and tools into three classes; user interface toolkits, user interface management systems (UIMSs), and Specification Techniques.

3.5 User Interface Development Systems and Tools

    3.5.1 User Interface Toolkits

        3.5.1.1 HP Toolkit

        3.5.1.2 Macintosh Toolbox

        3.5.1.3 Others

    3.5.2 User Interface Management Systems

        3.5.2.1 Presentation Management

        3.5.2.2 Behavior Management

        3.5.2.3 Automated Layout

    3.5.3 Specification Techniques

        3.5.2.1 Grammar-based Formalisms

            3.5.2.1.1 Command Language Grammar

            3.5.2.1.2 Backus-Naur Formalism

        3.5.2.2 Network Formalisms

        3.5.4.3 State Transition Networks

        3.5.4.4 Petri Nets

        3.5.2.5 Graphical Specification

The final sub-branch of interface engineering refers to techniques that can enhance the interface engineering process, these include iterative design and rapid prototyping.

3.6 Interface Engineering Techniques

    3.6.1 Iterative Design

    3.6.2 Rapid Prototyping

Human-Computer Interaction Applications

At present there is no standard user interface that can be used across all applications. The type of application has a major impact on the way that the user interface is constructed. For instance, real-time applications have to utilize methods for getting critical time-dependent information to the user or operator. Thus at a detailed level, much of the information and techniques in HCI will be application-specific, necessitating the incorporation of applications within the hierarchy. Areas that have received attention from researchers in human-computer interaction include:

4.1 Real Time Applications

4.2 Information Technology

4.3 Advanced Programming

4.4 Manufacturing and Industry

4.5 Computer Assisted Learning

(Continued)

TABLE 4.1
(Continued)

Further classification is shown below, however it is recognized that this portion of the taxonomy is relatively incomplete.

4.1 Real Time Applications
    4.1.1 Process Control
    4.1.2 Satellite Monitoring
    4.1.3 Video Games
    4.1.4 Military Applications
4.2 Information Technology
    4.2.1 Management Information Systems
    4.2.2 Decision Support Systems
    4.2.3 Database Management Systems
        4.2.3.1 Relational Databases
        4.2.3.2 Object-Oriented Databases
    4.2.4 Expert Systems and Intelligent Databases
        4.2.4.3 Data Analysis
        4.2.4.4 Consultation
    4.2.5 Information Retrieval
        4.2.5.1 Querying
        4.2.5.2 Browsing
    4.2.6 Learning Support Environments
    4.2.7 Text Editing
    4.2.8 Office Information Systems
    4.2.9 Simulation
        4.2.9.1 Discrete-Event Simulation
        4.2.9.2 Continuous Systems Simulation

In the classification of information technology above we have outlined four subcategories of information seeking. We include data analysis as information seeking here because it is a method of extracting and obtaining information. It may involve a number of summarization methods including statistics, operations research, time series analysis, and graphical presentations. Consultation refers to the process of getting information from an advisory source such as an expert system or a decision support system.

4.3 Advanced Programming
    4.3.1 Visual Programming
    4.3.2 Knowledge Engineering and Logic Programming
    4.3.3 Computer-Aided Software Engineering (CASE)
    4.3.4 Simulation
4.4 Manufacturing and Industry
    4.4.1 Computer-Aided Design (CAD)
    4.4.2 Computer-Integrated Manufacturing (CIM)
    4.4.3 Process Planning
    4.4.4 Inventory Management
    4.4.5 Transportation and Shipping
    4.4.6 Accounting and Billing
    4.4.7 Payroll Systems
4.5 Computer Assisted Learning
    4.5.1 CAI (computer-assisted instruction)
    4.5.2 Intelligent Tutoring Systems and Intelligent CAI
    4.5.3 Learning Support Environments

Note. From Chignell (1990). Reprinted with permission.

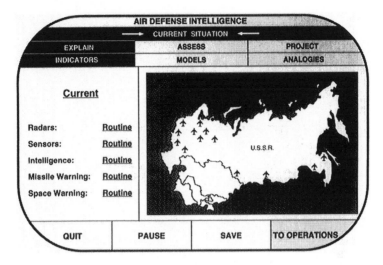

FIG. 4.1. Air defense storyboard.

Figures 4.1 through 4.5 suggest how we went about this initial case study. It was undertaken in order to assess the viability of the relationship between cognitive science and user–computer interface design and development. The displays indicate how we "matched" cognitive processes (and what we know about them) and display and interaction features.

They suggested that there was indeed a strong link between the field and expected enhancements in human–computer interaction and performance. But we did not conduct any tests of the displays or interaction routines. Our

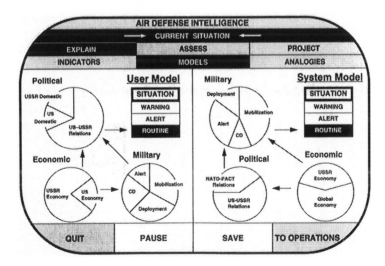

FIG. 4.2. Air defense storyboard.

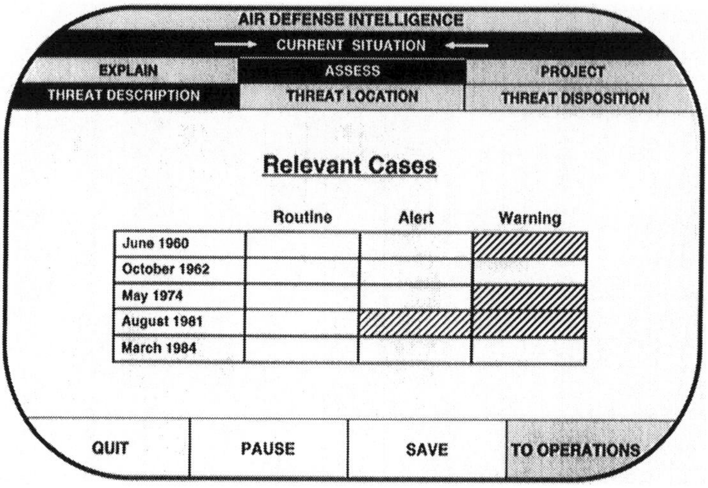

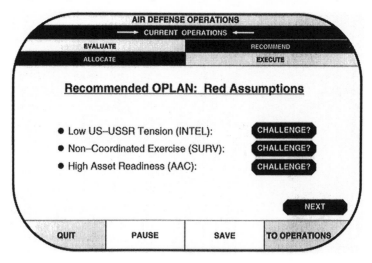

FIG. 4.3. Air defense storyboard.

assumption was, however, that tests would reveal enhanced performance, although we were unable to ever acquire "baseline" data on the current interface.

The Weapons Direction Case Study

This case study identifies the requirements connected with military air traffic control. It describes four major feature changes—inspired by what we know about cognitive problem solving, workload, situation assessment, and atten-

FIG. 4.4. Air defense storyboard.

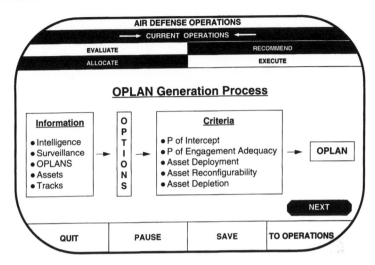

FIG. 4.5. Air defense storyboard.

tion—that include the addition of color (to enhance attention), the addition of high-threat symbology (to improve attention and assessment), the addition of an on-screen menu (to improve attention and reduce workload), and the integration of a quasi-automated problem-solving process (to reduce workload).

The case study describes experiments designed to test the impact of these changes. Although the results were not overwhelmingly positive, they clearly point in the direction of impact.

Expert System Interface Assessment

This case study reports on an experiment that investigated the effect of different real-time expert system interfaces on operators' cognitive processes and performance. The results supported the principle that a real-time expert system's interface should focus operators' attention to where it is required most. However, following this principle resulted in unanticipated consequences. In particular, it led to inferior performance for less critical, yet important cases requiring operators' attention. For such cases, operators performed better with an interface that let them select where they wanted to focus their attention. Having a rule generation capability improved performance with all interfaces, but less than hypothesized. In all cases, performance with different interfaces and a rule generation capability was explained by their effect on cognitive process measures.

The Order Effects Case Study

This case study reports on some experiments designed to investigate belief-updating models developed by Einhorn and Hogarth (1987) and Hogarth and Einhorn (1992). The model predicts that when information is presented sequentially and a probability estimate is obtained after each piece of information, people will anchor on the current position and then adjust their belief on the basis of how strongly the new information confirms or disconfirms the current position. Moreover, they hypothesized that the larger the anchor, the greater will be the impact of the same piece of disconfirming information. Conversely, the smaller the anchor, the greater will be the impact of the same piece of confirming information. This differential weighing of the most recent piece of information is predicted to result in different anchors from which the adjustment process proceeds and, in turn, different final probabilities based on the sequential order of the same confirming and disconfirming information.

The Submarine Display Redesign Case Study

This case study is based on many interactive, overlapping findings from cognitive science concerned with attention and memory; situational awareness; the way humans store, process, and retrieve knowledge; and other related issues. Not surprisingly, the overriding thesis of this case study is that if cognitive findings are used to develop a user interface, user performance will be improved. The domain was submarine weapons and launch control. A new interface was tested with reference to a baseline interface.

The purpose of this case study is to demonstrate that cognitively engineered user interfaces will enhance user–computer performance. This was accomplished by designing an interface using empirical findings from the cognitive literature. The interface was then tested and evaluated via a storyboard prototype. The cognitively engineered interface enhanced performance dramatically.

Displays and Interaction Routines for Enhanced Weapons Direction

This chapter describes research focusing on the design, development, prototyping, and evaluation of changes to the interface and interaction routines that weapons directors experience as they manage military air traffic in peacetime and during conditions of high threat, short time, and surprise.

The challenge was to identify, define, and validate a set of requirements that together comprise the weapons direction process. We undertook a detailed analysis of the weapons direction (WD) process and prioritized a set of requirements that we felt would enhance human performance.

REQUIREMENTS ANALYSIS

The requirements analysis process involved several steps. First, we studied airborne warning and command systems (AWACS) WD doctrine and practice. We also studied previous efforts to dissect the WD process via the development of IDEF charts. In addition, we conducted a series of interviews of weapons directors to determine the most complex tasks they perform and to gain additional insight into the nature and purpose of the weapons direction process.

CONVENTIONAL TASK ANALYSIS

We undertook a conventional task analysis via weapons direction handbooks and AWACS systems manuals supplemented with insight from domain experts.

The interviews, doctrine, scenarios and IDEF diagrams all provided insight into the nature of weapons direction. We distilled this insight into a set of requirements concepts and hypotheses—all detailed next.

107

INTERVIEW DERIVED HYPOTHESES

The following requirements concepts and hypotheses were derived from the interviews conducted with weapons directors:

- System cues to acknowledge appearance of new target, a newly detected target via sound cues, color changes, and/or flashing symbology.
- System suggestions or recommendations about which fighters to commit to target(s) and how to vector via color change, designation, or blinking; calculation of optimal interceptors and paths and graphic, animated displays of "candidates" based on the following criteria or attributes of the overall situation:
  - Amount of threat
  - Number and type of available assets
  - Combat Air Patrol (CAP) location/proximity
  - Other location activity
  - Target size or number
  - Range
  - Rules of Engagement (ROEs)
- System weighing of criteria; the results of the rank ordering to be displayed to WD with explanations. This capability will permit the development of intercept heuristics, which can be used to help an overloaded WD or check (via window-displayed constraint assessment) WD decision making (e.g., when a target is too close for an interceptor to optimize weapons systems).
- System monitoring and "control" of areas of responsibility between or among Senior Director via on-screen communication or display of targeted adversaries, committed interceptors, and the like.
- System listing, based on procedural, doctrinal, or situational data-driven heuristics, of the kinds of information that the WD should pass to interceptors in forms such as:
  - Likely threats
  - Location of supporting assets (such as tankers)
  - Likely radar activity

  calculated on a time line to maximize optimal communications between fighters and WD/AWACS based on empirical limitations of fighter radar.
- System versus WD model of what is critical information for fighter and what is extraneous via displays that "prod" WD or "remind" WD.
- Color shading of areas with high, medium, or low air or ground threats to better advise about where to go and where not to go.
- Special tagging of friendlies and communication of friendly locations to interceptors.

- System declutter functions, based on volume of traffic, targets, or threats, to clear screen of all but the highest threats (based on calculations of lethality, distance, P(k), etc.).
- System accumulation of WD events and actions and then a listing of items, queries, or problems to be communicated to ground via summary lists, queries, and so on; "call back" functions; a ring until contact is made; electronic message transfer of cumulative events/problems/queries; use of simple e-mail instead of all voice communications.
- System status monitors to check on conditions of all aircraft (e.g., fuel, time left in air, etc.) via active symbols for—via clicking—getting status data (except when critical stages were reached, then system would "interrupt" WD to tell him/her what's about to happen).
- Use of relevant libraries of intercept and tanker tracks as system-generated overlays of detect, commit, or intercept process.
- Criteria-based system-generated and WD-critiqued decisions, actions, and so on.
- "Mode" switching guidance, based primarily on radar coverage range, via system "reminders," color or symbol cues to WD, that range is shifting from AWACS optimal to fighter optimal; cues to alert of pending mode switching via window graphics of switch from AWACS to A10, F15, F16, and so on.
- System-guided inferences about interceptor and target behavior or tracks; embedded models of expected track behavior juxtaposed on WD's inference making via system agreeing or disagreeing with WD inference (independent of voice communication, which is confirmatory).
- Display declutter option via queries or availability of varying degrees of detail:

> | Declutter? | Click - Y
> | Full | Partial | Cancel |

- Capability to go to varying degrees of automation.
- Toggle switch controls that permit decluttering but restrict complete decluttering if the threat is high or as situation changes significantly.
- Prioritization of situation assessment data in composite format; redrawn or conversion of data for WD, fighter(s), or C2; geographic window in or on digitized map screen (current AWACS display).
- System-generated hypotheses about optimal or hypothetical allocation of fighter assets based on embedded system models of expected behavior and likely or unlikely if–then conditions; capability to respond to WD-initiated hypothesis testing regarding asset allocation based on changing threats, pace, and so on.

- System-generated attribute-driven recommendations about intercept plan via displays to WDs of allocation or intercept plan graphically, by tagging assets by color, blinking, or encircling.
- Encircling operation via one gesture, to simplify hooking, unlike two hooks (per two symbols) that were much like the Mac's lasso capabilities; augmentation with coding of lassoed area.
- System change of basic displays; symbols to icons of aircraft types (Bears, Blackjacks, Migs, F15s, F16s, A10s, etc.) to determine impact on friend or foe differentiation.
- System status monitors and calculations.
- Procedures for dealing with multiple track situations versus one or two track situations via system assuming greater and greater workload by generating and displaying hypotheses, checking WD options generation process, and so on.
- Situational profiles embedded in system to trigger overall UCI procedures or processes; for example, "long range," "close control," "few tracks," "two fighters" (vs., e.g., short/loose/many and 10) via adaptive control, system-task allocation, and the like; system recognition of situations via situational attributes; experimentation with whole new display concepts (decluttering, adaptive symbology, color coding, navigational cues, and the like).
- AWACS, which is best suited for: long range, close control, few tracks, and few fighters.
- Improvement, presumably, for the mission in converse situational performance.
- Procedural template that should appear to remind the WD of what must be done, has been done, and so on, as a helping hand during complex rapidly paced situations; cubes or boxes could be used to designate steps and as they are completed they would change form or colors; such navigational aids would appear more often as a function of situational complexity, for example:

Tasks

| Commit | Allocate Assets | Monitor; Modify | Communicate |

Considerations

| Safety | Complexity | ROE |

These and/or other options could be activated—with submenus—or "filled" as they were performed. Safety, for example, would have heuristics to guide the WD, available data procedures regarding, for example, International Friend or Foe (IFF) procedures, altitude, radar cut data, and the like:

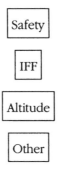

"If emissions are not confirmed by two sources, suspend kill orders . . ."

- Embedded model of pilot decision-making process to give the WD a pilot's eye of the situation and his or her mission, via displays of pilot's list of necessary data and what on the list has or has not been received and transmitted (and when):

Last Update: 1400
Pilot View - F16 - A

Track Info Threat Status
Vector Data

Exact Heading
Exact Altitude
Exact Speed

This is suggested because the WD needs to communicate extensively with pilots and because of the pilots' need for this information. Color can be used to designate information that has or has not been transmitted.

- Same process for ground controllers; template for communication.
- Awareness of warning level; awareness of complexity; awareness of ROEs and other constraints.
- Tabular to graphic displays on bandits (and other situational data) using "normal" human factors principles.
- Fuel (actual and calculated) displays, which is input from pilot and then checked with system algorithms to determine general accuracy (user hooks asset, enters fuel, and system reacts); heavy reliance on pilot's judgment, as input to future calculations.

- Capability to generate checklists of things to do, what has been done, what needs more data, and so on.
- System performance of basic "data cleaning" and labeling tasks; care and feeding of the CSM process itself is very time-consuming.
- System adaptation of more time for computer system maintenance (CSM) as complexity grows.
- Keyboard versus toggles.
- Three levels of problems and opportunities:
  - Screen interpretations
  - Allocation/matching problems
  - Time management problems (especially under complexity)
- Commitment to intercept:
  - System prioritization of targets or threats via automation
  - Automation of targets or weapons pairing process
  - Auto constraint checking
  - Probabilistics
  - Automation of WD/A2
- Development and implementation of a "shot clock" concept.
- Designation of inappropriate intercepts.
- Transitional processes from or to return to base (RTB), search and rescue (SAR), and so on, as well as other missions.
- Replacement of buttons with auto commands or on-screen or keyboard command options.
- Use of colors for correlation.
- Clean up symbology.
- Ranges associated with objects.
- Displays of feasible ranges.
- Improvement of data entry processes.
- Auto calculation of routes; analysis of routes.
- Fuel checks and "reminders."
- Use of colors, shading, and so on.
- System evaluation of feasibility of options.
- System facilitation of communication among team members.
- Color designation; for example, blue to designate water, brown for land, and so on.
- Icons of aircraft by type.
- Color coding of danger zones surface to air missile (SAM) circles to transparent red zones; anti aircraft artillery (AAA) and ocean-based threats, accordingly.

- Extensive use of windows for: reminders, checklists, and status-system control and communication.
- Online or stationary visual navigational and process cues.
- Embedded WD process model for "checking" or displaying WD actions.
- Online, stationary displays, icon, or color graphics of situational complexity, which are dynamically updated as complexity rises.
- Cues to user that situation is transitioning from one state of complexity to another—higher or lower—state of complexity based on complexity "drivers"; for example, number of tracks, intercept, or control.
- System signals of complexity transition via sound and updated displays, and the like, if number of tracks increases by 20%.
- Development of a "pilot perspective" display, perhaps a matrix of cells that represent tasks, updates (time-stamped), need-to-do items (identified by flashing).
- Access to previous exercise and actual cases pertinent to situation on hand, available as overlays to existing tracks and intercept solutions.
- Embedded inference models of expected red or adversary behavior about likely track behavior and likely outcome of intercept solutions, and so on.

IDEF-DERIVED HYPOTHESES

In addition to interview data, we relied on previous requirements modeling efforts. IDEF charts have been developed for the weapons direction process, and we analyzed them to identify important requirements, such as those that follow:

- IDEF charts reveal very high operator workload—far higher than necessary given computational or UCI and display technology.
- Major questions arise about system or operator task allocation; major opportunities arise from auto- and quasi-automation of tasks.
- Major questions arise about symbology: how it should be generated, displayed, developed, fidelity, and so on.
- Major questions arise about the hooking procedure and process; should we simulate improvements (via the "lasso," for example)?

The requirements concepts and hypotheses and the analysis of the empirical findings from cognitive science led to an emphasis on feature enhancements to increase situation awareness, reduce workload, and increase attention. A number of feature enhancements to the existing interface were thus developed and evaluated (see later) to determine the effect of introducing new interface features and interaction routines to the WDs.

These enhancements represented hypotheses about impact. We hypothesized that because of the high cognitive load of many of the WD tasks, reductions in such load—via increased situation awareness, increased attention, and an overall in task load—would enhance performance.

THE EXPERIMENT

This section is divided into five parts dealing with: (a) the experimental design and hypotheses, (b) the experimental testbed, (c) more detailed descriptions of the two system interfaces, (d) the experimental procedures, and (e) the dependent measures.

Design and Hypotheses

The original design was a 2 (interface) × 3 (experience level) factorial design. System interface was a within-subject variable; experience was a between-subject variable. As described next, failures of the basewide computer network at Brooks AFB, which was the site of the experiment, disrupted the counterbalancing of the order in which participants worked with the system interfaces. Consequently, a third factor—the order in which the participants worked with the two interfaces—was added to the design. "Order" was a between-subject variable.

The two system interfaces were (a) the current Airborne Warning and Control System (AWACS) interface, as implemented at the Aircrew Evaluation Sustained Operations Performance (AESOP) facility at Ft. Brooks, TX; and (b) an alternate interface, as guided by a cognitive systems engineering focus. The alternate interface represents the initial prototype for how the AWACS interface might be improved from a cognitive systems engineering perspective. Consistent with the literature (e.g., Adelman, 1992), the experiment represented an evaluation effort designed to obtain empirical and subjective data that could be used to assess the strengths and weaknesses of the initial prototype and, in turn, lead to its subsequent improvement.

Both interfaces were designed to permit AWACS Weapon Directors (WDs) to perform multiple missions. The experiment, however, focused solely on the Defensive Counter Air (DCA) mission, principally pairing friendly air defense fighters (ADFs) against enemy aircraft so that the latter would be intercepted and destroyed prior to successfully attacking friendly ground assets.

The AESOP facility and the system interfaces are described later. Here we identify, at a conceptual level, (a) how the alternate system interface differed from the current system interface, and (b) what cognitive processes were hypothesized to be improved by the change. The assumption throughout was that interface changes that improved cognitive processing would, in turn, improve overall performance on outcome measures.

The four principal changes and predicted effects were:

1. Color was added to the system interface. This was predicted to improve two cognitive processes: primarily, the WDs' attention (in terms of enhanced situation awareness) and, secondarily, WDs' situation assessment (in terms of judgments about future events).
2. High-Threat Symbols were added to improve primarily attention, but also situation assessment.
3. An On-Screen Menu was added to replace mechanical switches on a side panel of the interface and, thereby, improve attention because WDs could keep their eyes on the screen. In addition, the middle button on the three-button track ball permitted WDs to return from the on-screen menu to their previous place on the screen, which was marked by a large cross-hair forming a "t" on the screen. It was hypothesized that the on-screen menu and automatic return capability also would reduce the WDs' workload level.
4. A Quasi-Automated Nomination (QAN) capability was added to support decision making by using simple heuristics considering speed, location, direction, and fuel to recommend initial pairings of ADFs against enemy aircraft.

Eighteen WDs from Tinker AFB, OK, participated in the experiment. The WDs participated in crews of three, but the experiment was designed so that each WD had the same number of air threats and ADFs, and so that no threats crossed WDs' control boundaries. Consequently, we could collect and analyze the data for each WD separately.

All participants were certified AWACS WDs, but they differed in their flight experience. Three levels of an experience factor were created by distinguishing between WDs with (a) less than (or equal to) 500 hours, (b) between 500 and 1,000 hours, and (c) greater than (or equal to) 1,000 hours of AWACS flight experience. There were six participants for each level of experience.

An additional six WDs participated, but their data could not be used because basewide, computer system failures at Brooks AFB disrupted the experimental sessions for two crews. These system failures also disrupted counterbalancing of two additional factors: the time of day each system interface was tested, and the type of scenario. We counterbalanced both factors, that is, had half the sessions with the alternate system interface conducted in the morning and half in the afternoon of the test day, and had half with one scenario and half with another scenario, in an effort to control for extraneous factors that might affect performance. The basewide, system disruptions resulted in 12 of the 18 participants using the alternate system in the morning and 6 in the afternoon. Moreover, participants who used the alternate system in the morning were more likely to have used one scenario than another.

The order in which the two system interfaces were used was added to the experimental design in an effort to assess statistically whether it had any effect on system performance. Level 1 of the "order" variable was when participants used the alternate system interface first (i.e., in the morning) and the current interface second (i.e., in the afternoon). Level 2 was when participants used the current interface first and the alternate interface second.

From a statistical perspective, there were only six (instead of eight) participants in each cell of the original 2 (interface) × 3 (experience) design. This reduced the statistical power desired for testing interface × experience interactions, as well as for the two main effects. This is important because previous research conducted using the AWACS simulator in the AESOP facility had found large individual differences. Although the unit of analysis in that research had been groups, not individual WDs, we also expected considerable individual differences on a number of our measures. The reduced sample size reduced our ability to estimate the means and standard deviations for these measures and, in turn, the power of the statistical tests. However, statistically significant results are particularly noteworthy because they were obtained when it was harder to do so.

Experimental Testbed

The experiment was conducted during June and July 1992 at the AESOP facility. The facility is described in detail in Schiflett, Strome, Eddy, and Dalrymple (1990). The AESOP facility has four crew stations configured as AWACS WD consoles:

> These consoles have high resolution graphics displays, modular switch panels with programmable switch function, communication panels, QWERTY keyboards, and trackballs. Several high fidelity, low resolution video terminals serve as consoles for simulation pilots, ground controllers, and investigators. The AESOP computer systems consist of: (1) A cluster of 2 VAX 11/780s, 2 MicroVAX IIIs, and a VAXstation III/GPX; (2) [f]our high resolution, color graphics Sillicon Graphics 4D/50 workstations, and (3) [m]ultiple disk drives, tape drives, and printers. A 10-node communication network provides audio communication during simulations. (p. 4)

The Two System Interfaces

The interfaces consisted of the current interface for the system in use (and on which personnel trained and the "enhanced" interface), redesigned from requirements and findings from cognitive science.

We adopted a storyboard prototyping approach to the design, development, and evaluation process (Andriole, 1990, 1992). This approach permitted the design and development of screen displays that were inspected for requirements traceability and then were implemented into the software.

Screen displays from the process appear in Figures 5.1 through 5.16. Note how the displays evolved over time, as they were subjected to requirements traceability and likely impact criteria. The final figures represent selections from the displays that actually appeared in the enhanced interface.

The screens also suggest how individual features can be added or deleted based on judgments about what is likely to succeed (and fail).

Experimental Procedures

Each participant spent two days at the AESOP facility. The first day was for training the participants in using both systems. Training always began with the current system interface, which was quite similar to the actual AWACS system except for three differences: (a) the use of a standard QWERTY alphanumeric keyboard instead of the AWACS crewstation keyboard, (b) the use of a smaller trackball for "hooking" aircraft, and (c) the availability of only the most commonly used functions. These differences were discussed with participants after introductory remarks on the first day, and they were given an opportunity to practice using the current system prior to discussion of the alternate system interface.

Training for the alternate system interface had three principal stages. First, members of the research team described the four principal differences between the two system interfaces by using transparencies and paper representations of the alternate interface. Second, after participants understood these changes, they were given an opportunity to practice using the new system interface. The practice focused primarily on the step-by-step use of the on-screen menu and quasi-automated nomination procedure. This practice typically took the remainder of the morning session.

Third, participants used the alternate system interface for a 3-hour scenario in the afternoon of the first day. This scenario was comparable to the test scenarios in all respects except that it was not as difficult. Participants were repeatedly told to practice using all the features of the new system, but they often failed to do so because they became concerned about their performance. In an effort to counteract this tendency, the practice scenario was paused for 10 minutes at its midpoint to discuss the features of the alternate system interface and to attempt to ensure that all participants knew how to use them. In addition, a half-hour discussion was held after the practice session to address problems participants had in using the alternate interface's features. Even so, there were clear differences in participants' ability to use the on-screen menus and, particularly, the quasi-automated nomination procedure after training.

Performance with both interfaces was assessed using 3-hour test scenarios during the second day. One interface was used in the morning, and one was used in the afternoon. (As discussed previously, the original experimental design counterbalanced the time of day and type of scenario for which each system was used, but basewide computer system failures resulted in the loss of data for two groups and disruption in the counterbalancing scheme.)

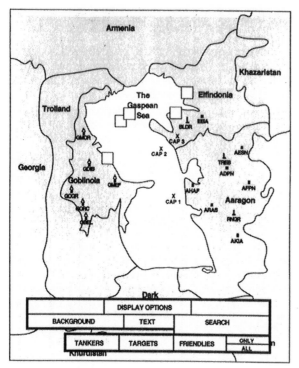

FIG. 5.1. Initial "storyboard concepts."

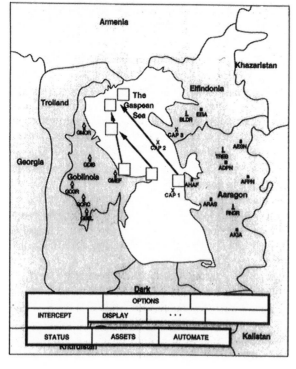

FIG. 5.2. Initial "storyboard concepts."

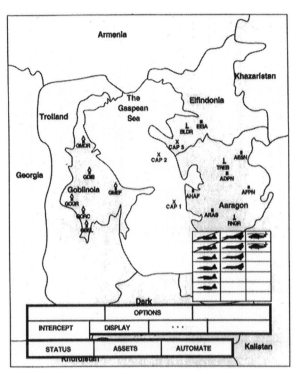

FIG. 5.3. Initial "storyboard concepts."

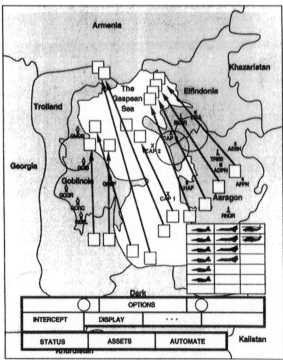

FIG. 5.4. Initial "storyboard concepts."

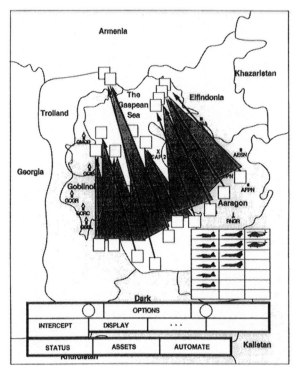

FIG. 5.5. Initial "storyboard concepts."

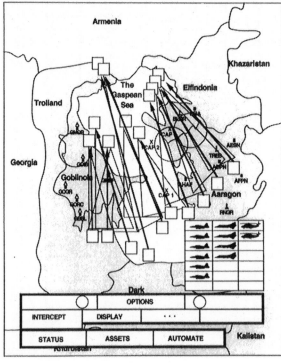

FIG. 5.6. Initial "storyboard concepts."

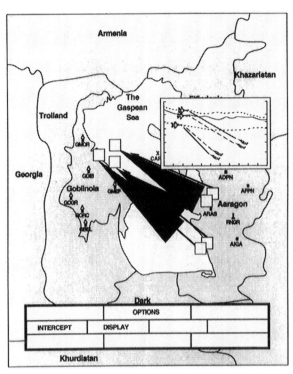

FIG. 5.7. Initial "storyboard concepts."

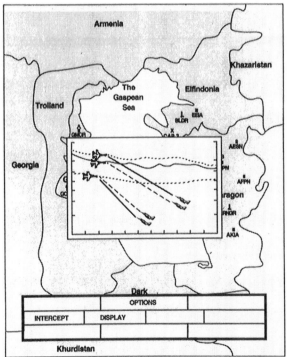

FIG. 5.8. Initial "storyboard concepts."

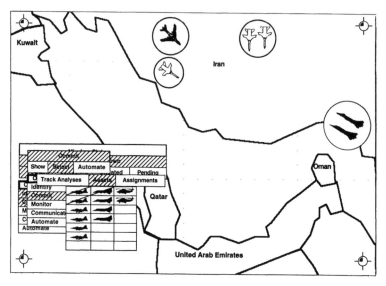

FIG. 5.9. Initial "storyboard concepts."

Participants were given a half-hour practice session on the system prior to its use for the test scenario. After completion of each test scenario, participants completed the subjective workload instruments described later.

DEPENDENT MEASURES

The hypothesis at its most general level was that the alternate system interface would improve cognitive functioning and, in turn, performance. Numerous

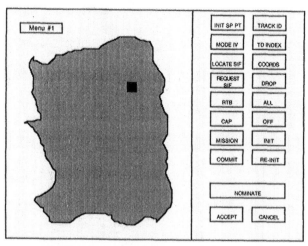

FIG. 5.10. This storyboard represents the display at rest. The on-screen menu is at the right.

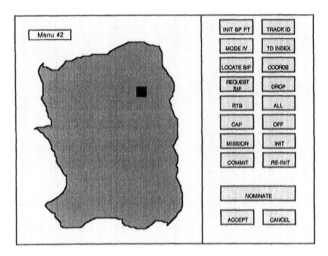

FIG. 5.11. This storyboard is a repeat of Fig. 5.10, but the menu is now active. At this point the WD would select the desired action by moving the menu cursor onto the desired "button" and pressing the hook button on the trackball. By depressing the middle trackball button, the cursor would return to its previous point, and the menu would deactivate. If the WD "rolls" the cursor out of the menu, the cursor returns to the conventional square representation and the two intersecting lines remain on the screen for 5 seconds. This allows the WD to return to the previous place on the screen and does not penalize him/her for what could have been a mistake (rolling out).

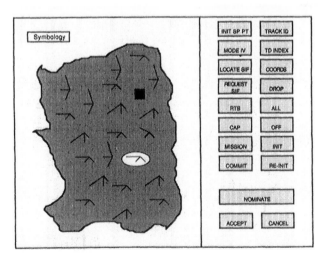

FIG. 5.12. This storyboard represents the system displaying enemy tracks. The track that is encircled represents a high-threat track. This track is either a jammer or a high-fast flyer.

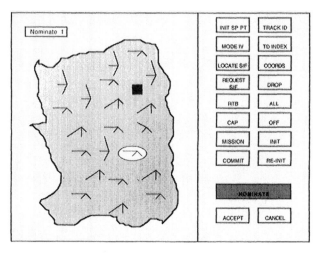

FIG. 5.13. Basic reconfigured AWACS display screen.

process measures were developed to assess the interfaces' effect on various cognitive processes. In addition, a number of outcome measures were assessed because they are typically used to measure WD performance. The cognitive process measures and outcome measures are described in turn.

Cognitive Process Measures

Measures were developed to assess five cognitive processes: attention, workload, memory, situation assessment (in terms of judgments about future events, not awareness of the current situation), and decision making. We make

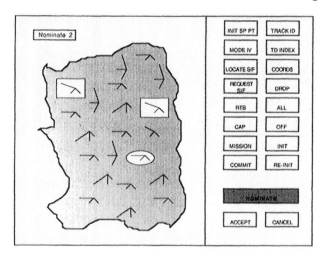

FIG. 5.14. Reconfigured display with active menu.

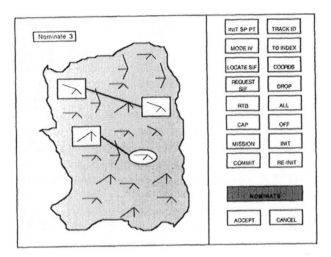

FIG. 5.15. This storyboard represents the system's suggestions. The system has placed squares around the friendly assets, squares around the enemy targets, and intercept lines to show the pairings. At this point the WD can Accept any or all of the pairings, Cancel any or all of the pairings, or monitor the situation until a decision can be made. If the WD Accepts the suggested pairings, the system executes a Commit switch action for the pairing. To Accept, the WD simply moves the cursor to either of the tracks in the desired pairing and presses the hook button on the trackball. The system then executes the Commit switch action and provides the WD with the necessary intercept information. To Cancel selected tracks, the WD simply selects Cancel from the menu, and moves the cursor to either of the tracks in the pairing to be cancelled and presses hook on the trackball. To cancel all pairings, the WD simply "double-clicks," using the hook button on the trackball, on Cancel from the menu.

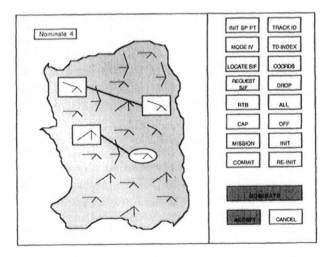

FIG. 5.16. Reconfigured display with "nominate options."

four points for introductory purposes before describing these measures. First, we distinguished between objective and subjective measures. Objective measures were based on data captured by the AESOP facility's computer system while participants performed the task; subjective measures were based on the participants' judgments.

Second, we had objective and subjective measures for all cognitive processes except memory, for which we developed only a subjective measure. We did not consider it cost effective to develop objective measures for memory because it was not a focus of our hypotheses.

Third, we developed a subjective questionnaire to capture the participants' thoughts about the effect of each of the four principal changes: colors, high-threat symbols, on-screen menu, and the quasi-automated nomination (QAN) capability. Most of the questionnaire focused on measuring the effect of each of these four changes on each of the five cognitive processes. These questions are described in this section briefly. The remaining questions measured whether the change had the desired effect envisioned by the research team. These questions are described in the subsection entitled "Subjective Questionnaire," which follows the subsection describing the outcome measures.

Fourth, in many cases, it was difficult to develop measures for one process that did not also, at least conceptually, measure another. This was particularly the case for distinguishing between (a) attention and workload, and (b) situation assessment and decision making. Said differently, it can be argued that some of our objective measures for assessing attention also measured workload, for effects on the former could also affect the latter, and vice versa. And for the Defensive Counter Air (DCA) mission performed by AWACS WDs, it can be argued that some of the objective measures developed to assess situation assessment also assessed decision making to a degree. Therefore, some measures are listed for two cognitive processes next.

Attention. Seven objective measures and one subjective measure were developed that could, in part, assess system effects on participants' attention. These are described briefly next:

• Two measures assessed participants' responses to 10 alerts presented visually by the AWACS display. One measure was the number of correct responses. A correct response was obtained by turning off the alert and being at the correct expansion level to examine the cause of the alert within 60 sec. Sixty seconds provided ample time to respond to the alert, even if a participant was at a different expansion level. (The expansion level determines the amount of air space presented on the AWACS display.) The second measure was the average response time for participants' responses (whether correct or incorrect), where 60 sec was used for those cases with no response. The hypothesis was that the alternate interface's use of color, high threat symbols, and on-screen menu would permit the participating WDs to attend better to visual stimuli.

- Two measures assessed participants' response to 11 questions asked by the Senior Weapon Director (SWD) via audio communications. (The SWD was a domain expert simulating that role for the purposes of the study.) One measure was the number of questions answered correctly within a 15-sec response window. The SWD recorded whether the question was answered correctly or not immediately after it was given by the participant. The second measure was the average response time for participants' responses (whether correct or incorrect), where 15 sec was used for those cases with no response. The hypothesis was that the alternate interface would reduce workload, thereby providing participants with the ability to (a) better attend to the situation, and (b) respond more accurately and quickly to aural stimuli with the alternate than current interface.

- The percentage of time participants spent at the larger expansion levels (i.e., < × 8) was assessed. The hypothesis was that the alternate interface's color and symbols would permit the participants to spend more time at larger expansion levels where they could see greater volumes of air space and, thereby, better attend to aspects of the situation. (Percentage was considered more informative than knowing the amount of time at larger expansion levels.)

- There were three objective process measures dealing with recorrelating symbology, that is, putting track symbols back on radar blips that had become uncorrelated due to aircraft maneuvering. The three measures were (a) time to recorrelate tracks, (b) percentage of incorrect recorrelations (i.e., putting track symbols on the wrong radar blips), and (c) the average amount of time symbols were incorrectly correlated. The hypothesis was that the alternate system's features would permit the WDs to attend to the situation better and, thus, more effectively deal with incorrectly correlated tracks. (Percentage of incorrect recorrelations was used to control for the differing number of uncorrelated symbols and radar blips caused by different aircraft maneuvers.)

- Participants' subjective assessment of the alternate system interface's effect on "attention" and its effect on "situation awareness," which is often used synonymously to describe the WD's state of attention. Participants answered four questions about the effect of the alternate system on their ability to focus their attention and four questions about its effect on their situation awareness. One question was asked for each of the four "principal changes" to the interface: color, high-threat symbols, on-screen menu, and the quasi-automated nomination (QAN) capability. Participants responded using a 5-point scale going from "Much Easier" to "Much More Difficult." The question measuring the effect of colors on attention follows for illustrative purposes:

The colors affected my ability to better focus my attention (as needed):				
Much Easier	*Somewhat Easier*	*No Effect*	*Somewhat More Difficult*	*Much More Difficult*
—	—	—	—	—

Workload. Seven objective measures and three subjective measures were developed that could, in part, assess interface effects on participants' workload: These are described briefly as follows:

- Accuracy and speed in responding to visual stimuli. In addition to enhancing attention, it was hypothesized that the on-screen menu would reduce the level of workload, as compared to the current system's separate mechanical switch panel, thereby permitting more accurate and faster responses to visual stimuli.

- Accuracy and speed in responding to aural stimuli. Similarly, it was hypothesized that the alternate system would permit more accurate and faster responses to aural stimuli.

- Percentage of commits with Stern or Stern Conversion (S or V) specified. When ADFs are committed against enemy aircraft, the WD is supposed to specify whether the intercept is to be by S or V for ground personnel keeping track of the air battle. This specification requires a separate switch action. It was hypothesized that it would require less work to perform this switch action with the on-screen menu than with the mechanical switch panel. As a result, we hypothesized that participants would do it more often with the alternate than current system. (Percentage was used to control for possible differences in the number of committed aircraft.)

- Average time to recorrelate tracks; percentage of incorrect recorrelations and the average amount of time tracks are incorrectly correlated. It was hypothesized that the reduced workload with the alternate system would give participants more time to deal with uncorrelated tracks than with the current system and, thus, result in faster recorrelations, fewer incorrect recorrelations, and faster correlations when errors occurred.

- The NASA-TLX (Task Loading Index) (Hart, 1988). This is a multidimensional rating technique for measuring subjective workload. Participants rated each of the two system interfaces on six dimensions: mental demand, physical demand, temporal demand, performance, effort, and frustration level. Ratings for each dimension were made on a scale using low and high as its endpoints and entered directly by participants via a computer program for implementing the NASA-TLX technique. In addition, participants responded to a series of paired comparisons of the six dimensions designed to calculate the weight (or subjective importance) of the dimensions in determining the overall level of workload for the DCA mission. An overall workload measure was calculated for each interface by (a) multiplying the participant's weight and rating for each dimension and, then, (b) summing the products.

- The Subjective Workload Dominance (SWORD) Technique (Vidulich, Ward, & Schueren, 1991). SWORD uses a series of paired comparison questions to assess the relative workload of performing different tasks with different system interfaces. In the current study, there were three tasks (recorrelating symbology, pairing ADFs, and conducting intercepts) and two interfaces (current and alternate). Participants were asked to compare the relative workload for each task–interface combination resulting in 15 comparisons. Then, using a software package provided by Vidulich, the geometric mean was calculated for each level of each of the two variables. The higher the geometric mean, the higher was the level of subjective workload. The geometric means were the cell entries to the Analysis of Variance (ANOVA) tests described in the results section for SWORD. In addition, the software package performed a consistency ratio calculation to determine the degree to which the participant's paired comparison responses were consistent with one another. Ratios greater than 0.10 were considered "inconsistent" for the study, as suggested by Vidulich.

- Participants' subjective assessment of the alternate system interface's effect on workload. Participants answered three questions about the effect of the alternate system on their workload. One question was asked for color, on-screen menu, and QAN. A question was not included for high-threat symbols due to an oversight when constructing the questionnaire. Participants responded using the 5-point scale going from "Much Easier" to "Much More Difficult" as described earlier.

Memory. The only measure of the alternate system's effect on participants' memory was the subjective measure generated by asking participants the effect of each of the four "principal changes" on their ability to remember important information. Participants responded using the 5-point scale previously described.

Situation Assessment. There were two objective measures and one subjective measure assessing situation assessment. Due to the nature of the DCA task, both situation assessment measures also measure the quality of the WDs' decision-making process. These measures are considered next.

- Ratio of Airborne Orders to Scrambles. It was hypothesized that the alternate system interface's color, high-threat symbols, and on-screen menu would not only improve WDs' attention, but their ability to assess future events and needs. One of the most critical events is when new ADFs are needed for the air battle. New ADFs might be needed because the number of enemy aircraft approaching friendly air space is larger than expected or to replace ADFs that either have been lost in combat or need to refuel or return to base to obtain weapons. It was reasoned that the more time that WDs give ground controllers to provide new ADFs, the better the WDs' situation assessment and decision making. Airborne orders give ground controllers more time than do scrambles to provide new ADFs and, more generally, to manage the air battle. Airborne orders were operationalized as ADF requests where the WDs said they needed the aircraft in more than 5 min; scrambles were requests where the ADFs were needed in less than 5 min.

• Percentage of Airborne Refuelings versus Return-to-Base Refuelings. It was hypothesized that the alternate system would better permit WDs to monitor the fuel levels of ADFs, to assess at what time different ADFs would have to refuel, and to plan for airborne refuelings. Airborne refuelings take less time, but require more coordination than having the aircraft return to base for fuel. (Percentages were used instead of ratios because some participants had 0 airborne refuelings, whereas other participants had 0 return-to-base refuelings; consequently, ratios would have resulted in zeros in the denominator.)

• Participants' subjective assessment of the alternate system interface's effect on situation assessment. Participants answered questions about the effect of each of the four principal changes on their situation assessment ability using the 5-point response scale described earlier.

Decision Making. There were four objective measures and one subjective measure of the quality of the WDs' decision-making process. Two of the objective measures—the percentage of airborne orders and airborne refuelings—also were used to measure situation assessment due to the close relation between situation assessment and decision-making processes for the DCA mission. Each measure is considered here:

• Ratio of Airborne Orders versus Scrambles. Airborne orders and scrambles represent decisions that the WDs make. Therefore, they not only measure situation assessment, they also measure aspects of the WDs' decision-making process. It was hypothesized that the alternate system would improve situation assessment and, in turn, decision making with the result being a higher percentage of airborne orders.

• Percentage of Airborne Refuelings to Return-to-Base. Similarly, WDs have to decide whether aircraft refuelings will be performed in the air or on the ground. This can be a difficult decision requiring WDs to balance whether the ADF still has an adequate array of weapons, the coordination requirements of airborne refuelings, the need for returning ADFs to the air battle as soon as possible, and the ease of sending ADFs back to their base for fuel and weapons. Therefore, this measure assesses the quality of the WDs' decision-making process as well as the quality of their situation assessment.

- Number of commits. It was hypothesized that the alternate system's Quasi-Automated Nomination (QAN) procedure would result in an increase in the number of hostile aircraft against which friendly ADF were committed.

- Percentage of missiles fired that missed. In addition to nominating what ADFs should be paired against what hostile aircraft, the QAN provides lines showing the pairings that can be used to represent the intercept trajectory. It was hypothesized that this feature, plus the features designed to improve the WDs' attention, would improve the WDs' decisions regarding the intercept trajectory. The better these intercept decisions were, the fewer the number of missiles that would miss their mark. (Percentages were used to control for differences in the number of commits and, in turn, the number of missiles fired against enemy aircraft.)

- Participants' subjective assessment of the alternate system interface's effect on decision making. Participants answered questions about the effect of each of the four principal changes on their decision making ability using the 5-point response scale described earlier.

We now turn to consider the outcome measures used to assess system effects on WD performance.

Outcome Measures of WD Performance

There were eight objective measures of WD performance. Again, the general hypothesis was that the alternate system interface would improve overall cognitive processing and, in turn, overall performance. Consequently, we did not specify individual cognitive process to outcome performance hypotheses. It should be noted from the outset that many of the performance measures are correlated, in some cases, highly correlated. However, many of the measures are used in actual settings and, therefore, are included herein.

Enemy Mission Accomplishment. A critical measure of WD performance is the extent to which the enemy accomplished their mission. Our hypothesis was that WDs would do better and the enemy worse when the WDs used the alternate system interface. We used four measures, each of which is considered here:

- Number of hostile air strikes completed. This is the ultimate measure of enemy mission accomplishment. (Note: Each participating WD had to deal with 20 hostile aircraft in each of the two scenarios.)

- Number of hostile penetrations of friendly airspace. The participating WDs were all certified and reasonably experienced. Therefore, we reasoned that there might be few hostile strikes completed, regardless of system features. However, it was quite possible that there would be significantly fewer penetrations of friendly air space with the alternate system interface.

- Average penetration distance for penetrators only. We hypothesized that if penetrations of friendly air space occurred, the WDs would be able to better anticipate these penetrations and recover faster with the alternate system interface.

- Average penetration distance for all hostile aircraft, where hostiles that fail to penetrate friendly air space were given a value of 0.

Enemy and Friendly Losses. These performance measures assess friendly as well as enemy performance, but their focus is on how well the enemy accomplished its mission. The next three measures assess friendly performance by focusing on enemy and friendly aircraft losses:

- The number of enemy aircraft destroyed by friendly ADFs.
- The friendly-to-enemy kill ratio, where the numerator is the number of friendly aircraft destroyed by enemy aircraft and the denominator is the number of enemy aircraft destroyed by friendly ADFs.
- Percentage of aircraft destroyed that were hostile.
- The total number of friendly losses, including losses due to hostile aircraft, enemy or friendly surface-to-air missiles, and running out of fuel.

Composite Measure. The domain expert who played the role of the Senior Weapons Director (SWD) during the study had previously developed a measure combining the above measures. Although the composite measure has not been validated using real-world data, it represented a method for obtaining an aggregate performance score. Consequently, we chose to use it as a dependent measure for the study. It is described verbally instead of using an equation format. Specifically, the composite score for a WD performing the DCA mission in the experiment was equal to:

- The negative square of the number of hostile strikes completed
- Plus the kill ratio times the number of hostile aircraft killed
- Minus the loss ratio (i.e., all friendly ADF losses) times the number of hostiles not killed
- Minus the kill ratio times the number of friendly ADFs lost (either due to enemy or friendly aircraft fire)

- Minus the friendly loss ratio times the number of friendly ADFs lost by fuel depletion or surface-to-air missiles (either friendly or enemy).

We now turn to consider the questions measuring whether the principal changes had the desired effect on a task performed by the WDs.

Subjective Questionnaire

Most of the questions in the subjective questionnaire attempted to measure the effect of the alternate system on cognitive processes. A number of questions, however, attempted to measure whether or not the principal changes had the desired effect on tasks performed by the WDs. These questions are considered, in turn, for each change. All questions used the 5-point response scale described earlier.

Color. It was hypothesized that the use of color would greatly assist WDs' ability to identify the boundaries and geographic features of the map. In addition, we asked whether the use of color helped WDs (a) locate high value targets, and (b) plan intercepts. It was hypothesized that color would have considerably less effect on these tasks.

High-Threat Symbols. It was hypothesized that the high-threat symbols would greatly assist the WDs in locating high-value targets. In addition, we asked whether the use of high-threat symbols helped WDs (a) identify boundaries and geographic features of the map, and (b) plan intercepts.

On-Screen Menu. It was hypothesized that the on-screen menu would greatly assist the WDs in executing critical functions (e.g., commit switch actions). Consequently, we asked the WDs this question.

QAN. It was hypothesized that the QAN would assist WDs in making their commit decisions. This hypothesis was based on the assumption that the rules used by the QAN generated credible solutions. Consequently, we asked the WDs this question.

RESULTS

We present the results for all objective measures, then the NASA-TLX and SWORD subjective workload measures and, last, the results of the subjective questionnaire.

Objective Measures

A 2 (interface) × 3 (experience) × 2 (order) ANOVA was performed for each objective process and outcome measure. The type of system interface was a within-subject variable; a participant's level of experience and the order in which they worked with the two system interfaces were between-subject variables.

It is important to note at the outset that results with $p < 0.10$, instead of the traditional $p < 0.05$ significance level, are presented because the less conservative significance level makes it easier to understand the more general reasons for system interactions across a number of process (and outcome) variables. However, we realize that a large number of statistical tests were performed and that, on the average, one would expect 10% of these tests to be significant at the $p = 0.10$ significance level by chance alone. With this concern in mind, we still chose the more liberal significance level in an effort to better understand what the data, in total, may be conveying about the performance of the alternate system interface. Our decision also was affected by the facts that (a) the statistical power of our tests was less than planned for, due to the loss of 25% (6 of 24) of the participants, and (b) the alternate system interface was an initial prototype. Subsequent revisions to the interface, if the government chooses to make them, need to be informed by the data as much as possible.

First, we present the results for the objective process measures and then the results for the objective outcome measures. Then we present a reanalysis of the data incorporating the participants' squadron as a (new) independent variable. This post-hoc analysis was justified by observation of the participants during the test sessions and visual examination of the data. As will be seen, squadron significantly affected performance with the two system interfaces.

Objective Process Measures. Table 5.1 presents the results of the ANOVAs for each of the interface and experience main effects and the interface by experience (I × E) interaction for each objective process measure. Table 5.2 presents the ANOVA results for the order main effects, the order by interface (O × I) interactions, the order by experience (O × E) interactions, and the order by interface by experience (O × I × E) interactions for the objective process measures.

Examination of Table 5.1 shows that there were only three significant main effects for the system interface and one significant interface by experience interaction:

- The current interface resulted in a faster response time to aural questions: $F(1,12) = 4.585$, $MSe = 468.222$, $p = 0.053$, current $M = 7.433$ sec, alternate $M = 8.367$ sec. (Note: M means mean.)

TABLE 5.1
Interface and Experience Results: Objective Process Measures

Cognitive Process Measured	Interface (I) Main Effect	Experience (E) Main Effect	I × E Int.
Attention and Workload			
- Accuracy: Visual Alerts	ns	ns	ns
- Response Time: Visual Alerts	ns	ns	ns
- Accuracy: Aural Questions	ns	ns	ns
- Response Time: Aural Questions	$p = 0.053$	ns	ns
- Time to Recorrelate	ns	ns	ns
- Percentage of Incorrect Recorrelations	ns	ns	ns
- Time Incorrectly Recorrelated	ns	ns	ns
Attention Only			
- Expansion Level	ns	ns	ns
Workload Only			
- Percentage of S or V Commits	$p = 0.028$	ns	$p = 0.040$
Situation Assessment and Decision Making			
- Ratio of Airborne Orders to Scrambles	ns	ns	ns
- Percentage of Airborne Refuelings vs. Return-to-Base	$p = 0.014$	ns	ns
Decision Making Only			
- Number of Commits	ns	ns	ns
- Percentage of Missiles that Missed	ns	ns	ns

- The current interface resulted in a larger percentage of S or V commits: $F(1,12) = 6.25$, $MSe = 0.002$, $p = 0.028$, current $M = 15.1\%$, alternate $M = 12.2\%$.

- The alternate system resulted in a larger percentage of airborne refuelings: $F(1, 12) = 8.271$, $MSe = .019$, $p = .014$, current $M = 18.6\%$, alternate $M = 33\%$.

- The alternate system resulted in (a) a much lower percentage of S or V commits for the least experienced participants (7.4% vs. 17%), (b) a slightly higher percentage for the most experienced participants (14.8% vs. 13.3%), and (c) about the same percentage for moderately experienced participants (14.4% vs. 14.8%): $F(2,12) = 4.263$, $MSe = .002$, $p = .04$.

In total, the results suggest that the alternate system had minimal effect on the process measures. In the few cases when it did have an effect, it was (a) negative with respect to attention and workload or depended on participants' experience level, and (b) positive with respect to situation assessment and decision making. Table 5.2 sheds additional light.

Examination of Table 5.2 shows that there were no order main effects, but a number of significant interactions. These are described briefly, in turn, as follows (the mean values for the interactions are presented in Table 5.3):

• Order × Experience Interaction for Response Accuracy to Visual Alerts: $F(2,12) = 8.999$, $MSe = 3.103$, $p = 0.004$. The most experienced participants did considerably worse, in terms of their average response accuracy for the two interfaces, when they worked with the alternate interface after working with the current system. In contrast, participants with medium experience did slightly better, on the average, when they used the alternate system interface after working with the current interface.

• Order × Experience Interaction for Response Time for Visual Alerts: $F(2,12) = 5.43$, $MSe = 7315$, $p = 0.021$. Again, the most experienced participants did considerably worse, in terms of their average response time for the two interfaces, when they worked with the alternate system after working with the current system. And participants with medium experience did better when they used the alternate interface after the current one.

• Order × Interface Interaction for Response Time for Aural Questions: $F(1,12) = 3.555$, $MSe = 468.222$, $p = 0.084$. There was a faster response time to aural questions with the alternate interface when it was used first and a faster time with the current interface when it was used first.

• Order × Interface × Experience Interaction for Response Time for Aural Questions: $F(2,12) = 3.645$, $MSe = 468.222$, $p = 0.058$. Examination of the data in Table 5.3 shows that order by interface interaction depends on the

TABLE 5.2
Order Main Effects and Interactions: Objective Process Measures

Cognitive Process Measured	Order (O) Main Effect	O × I Int.	O × E Int.	O × I × E Int.
Attention and Workload				
- Accuracy: Visual Alerts	ns	ns	$p = .004$	ns
- Response Time: Visual Alerts	ns	ns	$p = .021$	ns
- Accuracy: Aural Questions	ns	ns	ns	ns
- Response Time: Aural Questions	ns	$p = .084$	ns	$p = .058$
- Time to Recorrelate	ns	ns	ns	$p = .090$
- Percentage of Incorrect Recorrelations	ns	ns	ns	ns
- Time Incorrectly Recorrelated	ns	ns	ns	ns
Attention Only				
- Expansion Level	ns	ns	$p = .034$	ns
Workload Only				
- Percentage of S or V Commits	ns	ns	ns	ns
Situation Assessment and Decision Making				
- Ratio of Airborne Orders to Scrambles	ns	ns	$p = .069$	ns
- Percentage of Airborne Refuelings				
vs. Return-to-Base	ns	ns	ns	ns
Decision Making Only				
- Number of Commits	ns	ns	ns	ns
- Percentage of Missiles that Missed	ns	ns	ns	ns

TABLE 5.3
Cell Means for Significant Interactions Involving Order

Response Accuracy (number): Visual Alerts (Order × Experience)

| | Experience Level | | |
Order	Low	Medium	High
Alt 1st, cur 2nd	4.75	3.90	6.17
Alt 2nd, cur 1st	4.50	7.00	2.50

Response Time (in sec): Visual Alerts (Order × Experience)

| | Experience Level | | |
Order	Low	Medium	High
Alt 1st, cur 2nd	38.6	43.0	35.6
Alt 2nd, cur 1st	36.9	30.5	49.3

Response Time (in sec): Aural Questions (Order × Interface)

| | Interface | |
Order	Alternate	Current
Alt 1st, cur 2nd	7.917	7.792
Alt 2nd, cur 1st	9.267	6.717

Response Time (in sec): Aural Questions (Order × Interface × Experience)

| | Alt 1st, Cur 2nd Experience | | | | Alt 2nd, Cur 1st Experience | | |
System	L	M	H	System	L	M	H
Alternate	10.08	6.9	6.73	Alternate	4.60	13.7	10.9
Current	8.88	7.70	6.50	Current	6.00	5.70	7.53

Time to Recorrelate (in simulation cycles): Order × Interface × Experience

| | Alt 1st, Cur 2nd Experience | | | | Alt 2nd, Cur 1st Experience | | |
System	L	M	H	System	L	M	H
Alternate	3081	3245	1037	Alternate	3092	5290	11311
Current	1819	2594	3096	Current	4013	870	2804

Percentage of Time at Higher Expansion Levels: Order × Experience

| | Experience Level | | |
Order	Low	Medium	High
Alt 1st, cur 2nd	92.75	92.45	86.50
Alt 2nd, cur 1st	84.10	97.75	97.53

Ratio of Air Orders to Scrambles: Order × Experience

| | Experience Level | | |
Order	Low	Medium	High
Alt 1st, cur 2nd	.36	.29	.99
Alt 2nd, cur 1st	3.77	.23	.79

137

experience level too. When the alternate interface was used first, the least experienced participants responded faster with the current interface, whereas the participants with medium experience responded faster with the alternate interface. When the alternate system was presented second, just the opposite occurred.

• Order × Interface × Experience Interaction for Time to Recorrelate Tracks: $F(2,12) = 2.958$, $MSe = 9731799$, $p = 0.09$. When the alternate system was used first, the most experienced personnel were faster recorrelating the symbols on the tracks with the alternate rather than the current system interface. Just the opposite happened when the alternate interface was used second. Participants with a moderate level of experience were faster with the current interface, regardless of the order. (Note: The unit of measure for this process measure is simulation cycles, where one cycle occurs approximately every 2 sec depending on the number of symbols on the AWACS screen.)

• Order × Experience Interaction for Percentage of Time at the Higher Expansion Levels: $F(2,12) = 4.567$, $MSe = 0.006$, $p = 0.034$. On the average, the least experienced participants did better (i.e., spent more time at the larger expansion levels) when they worked with the alternate interface before working with the current interface. In contrast, both moderately and highly experienced participants did better, on the average, when they worked with the alternate interface after working with the current one.

• Order × Experience Interaction for Ratio of Airborne Orders to Scrambles: $F(2,12) = 3.365$, $MSe = 2.975$, $p = 0.069$. On the average, the least experienced personnel had higher ratios when they worked with the alternate interface second. In contrast, the average performance was better for the more experienced personnel when they worked with the alternate system interface first.

In total, these results suggest that the order in which participants worked with the alternate and current system interfaces did affect their performance on a number of process measures, depending on the participants' level of experience. The attention and workload measures were most affected; five of the nine measures had significant interactions. The situation assessment and decision-making measures were least affected; only one of four measures had significant interactions. In many cases where there was a significant effect involving experience, the most experienced personnel performed worse when they worked with the alternate system after working with the current system. This suggests that they may require more training with the alternate system than less experienced personnel to overcome carryover effects from the current system interface.

Objective Performance Measures.   There were no significant effects for any objective performance measure.

Analysis for Squadron. WDs from four AWACS squadrons participated in the experiment. Specifically, one WD participated from the 552nd squadron, seven participated from the 963rd, five participated from the 964th, and five participated from the 965th. Observations of the participants' performance during the test sessions and visual examination of the data raised concern that the participants' squadrons might affect performance with the two systems. Consequently, we reanalyzed the data incorporating the participants' squadron as a (new) independent variable. The one WD from the 552nd squadron was dropped from the analysis so that there would not be one participant in one level of the squadron variable.

The particular concern was whether squadron type significantly affected performance with the two interfaces; consequently, we used a 2 (Interface) × 3 (Squadron) design. Interface was a within-subject variable; squadron was a between-subject variable. We present the results first for the process measures and then the outcome measures. Because our concern was with whether there was a squadron effect, and not with how well or how poorly different squadrons performed, we do not present the scores for specific squadrons. Moreover, the reader should not assume that the order in which the squadrons are listed here matches the earlier order.

Table 5.4 presents the results of the reanalysis for the process measures. As can be seen, there is now only one significant main effect for system

TABLE 5.4
Interface and Squadron Results: Objective Process Measures

Cognitive Process Measured	Interface (I) Main Effect	Squadron (SQ) Main Effect	I × SQ Int.
Attention and Workload			
Accuracy: Visual Alerts	ns	ns	ns
Response Time: Visual Alerts	ns	ns	ns
Accuracy: Aural Questions	ns	ns	ns
Response Time: Aural Questions	ns	ns	$p = 0.098$
Time to Recorrelate	ns	ns	ns
Percentage of Incorrect Recorrelations	ns	ns	ns
Time Incorrectly Recorrelated	ns	ns	ns
Attention Only			
Expansion Level	ns	$p = 0.01$	ns
Workload Only			
Percentage of S or V Commits	ns	ns	ns
Situation Assessment and Decision Making			
Ratio of Airborne Orders to Scrambles	ns	ns	$p = 0.027$
Percentage of Airborne Refuelings vs. Return-to-Base	$p = 0.002$	ns	$p = 0.088$
Decision Making Only			
Number of Commits	ns	$p = 0.025$	ns
Percentage of Missiles that Missed	ns	ns	ns

interface. That main effect was for the percentage of airborne refuelings: $F(1,14) = 14.65$, $MSe = 0.009$, $p = 0.002$. Again, the percentage was higher with the alternate than current system interface (alt = 31.1%, cur = 19.7%; 17 participants). (Note: The previous main effects for response time for aural questions and the percentage of S or V commits are no longer significant due to some combination of the differences in the partitioning of the variance due to system and other variables, and/or differences in the degrees of freedom and number of participants.)

As shown in Table 5.4, there are two main effects for squadrons and three system-by-squadron interactions. Each is now considered briefly:

- Squadron Main Effect for Expansion Level: $F(2,14) = 6.51$, $MSe = 0.005$, $p = 0.01$. Squadron 1 spent the largest (mean) percentage of their time at the larger expansion levels (96.1%) than did squadron 2 (89.5%) or squadron 3 (86.3%).

- Squadron Main Effect for Number of Commits: $F(2,14) = 4.837$, $MSe = 130.35$, $p = 0.025$. Squadron 2 had more commits ($M = 47.1$) than did squadron 1 (35.22) or squadron 3 (32.1).

- System × Squadron Interaction for Response Time to Aural Questions: $F(2,14) = 2.76$, $MSe = 503.4$, $p = 0.098$. Squadron 1 did much better, on the average, with the current than alternate system (7.5 vs. 10.2 sec). Squadron 2 did slightly better with the current interface (5.7 vs. 6.98). In contrast, squadron 3 did better with the alternate system (7.68 vs. 5.94).

- System × Squadron Interaction for Ratio of Airborne Orders to Scrambles: $F(2,14) = 4.702$, $MSe = 0.259$, $p = 0.027$. On the average, squadron 1 had a higher ratio with the current interface (0.80 vs. 0.289). In contrast, both squadrons 2 and 3 had higher ratios with the alternate interface (0.358 vs. 1.138 and 1.506 vs. 1.60, respectively).

- System × Squadron Interaction for Percentage of Airborne Refueling: $F(2,14) = 2.913$, $MSe = 0.009$, $p = 0.088$. The percentage of airborne refuelings was higher for all three squadrons with the alternate system, which was why there was a significant main effect. However, the difference was much larger for squadron 2 (51.1% vs. 29.2%) than for squadron 3 (28.9% vs. 16.7%) or squadron 1 (18.3% vs. 15.1%).

Table 5.5 presents the results of the reanalysis for the outcome measures. As can be seen, there were significant system-by-squadron interactions for six of the eight outcome measures. As it will be remembered, there were no significant effects for any outcome measure using the original design.

Table 5.6 presents the mean values for each significant system-by-squadron interaction. Examination of these data shows that in all cases (a) participants from squadron 1 performed worse with the alternate interface, whereas (b) participants from squadrons 2 and 3 performed better with the alternate interface.

TABLE 5.5
Interface and Squadron Results: Objective Outcome Measures

Performance Measure	Interface (I) Main Effect	Squadron (SQ) Main Effect	I × SQ Int.
Enemy Mission Accomplishment			
Number of Hostile Strikes Completed	ns	ns	$p = 0.096$
Number of Hostile Penetrations	ns	ns	$p = 0.016$
Average Penetration	ns	ns	ns
Distance (Only for Penetrators)	ns	ns	ns
Average Penetration Distance			
(All Hostile)	ns	ns	$p = 0.02$
Hostile and Friendly Losses			
Number of Enemy Losses Due to			
Friendly ADFs	ns	ns	ns
Friendly to Enemy Kill Ratio	ns	ns	$p = 0.01$
Total Number of Friendly Losses	ns	ns	$p = 0.009$
Composite Score	ns	ns	$p = 0.047$

These results, in conjunction with those for the process measures, indicate that squadron type did significantly affect performance with the interfaces.

Subjective Measures

This section presents the results for the two subjective workload measures—NASA TLX and SWORD—and the subjective questionnaire eliciting the participants' opinions about the alternate interface in turn.

NASA TLX. Participants rated each of the two system interfaces on six dimensions: mental demand (MD), physical demand (PD), temporal demand (TD), effort (EF), performance (OP), and frustration level (FR). In addition, participants responded to a series of paired comparisons of the six dimensions in order to calculate the weight (or subjective importance) of the dimensions. An overall workload measure was calculated for each interface by (a) multiplying the participant's weight and rating for each dimension and, then, (b) summing the products.

Three 2 (Interface) × 6 (Dimension) ANOVAs were performed: one for the weights, one for the (unweighted) scores, and one for the (weighted) overall workload measure. Both interface and dimensions were within-subject variables because the participants completed the NASA TLX for each system interface. Each ANOVA is considered in turn.

We hypothesized that there would be a significant main effect for dimensions for the "weights" dependent variable. For example, we thought that Mental Demand would be more important than Physical Demand. We did

TABLE 5.6
F Ratios and Cell Means for Significant System × Squadron
Interactions for Objective Outcome Measures

System	Squadron		
	1	2	3
Number of Hostile Strikes Completed: $F(2,14) = 2.779$, $MSe = 0.914$, $p = 0.096$			
Alternate	0.286	0.80	1.00
Current	1.286	0	0.80
Number of Hostile Penetrators: $F(2,14) = 5.601$, $MSe = 7.422$, $p = 0.016$			
Alternate	10.3	5.0	8.6
Current	5.0	6.6	9.0
Average Penetration Distance (all enemy): $F(2,14) = 5.284$, $MSe = 70.32$, $p = 0.02$			
Alternate	22.55	7.05	16.64
Current	10.93	15.83	22.31
Friendly to Enemy Kill Ratio: $F(2,14) = 6.465$, $MSe = 0.009$, $p = 0.01$			
Alternate	.298	.240	.221
Current	.187	.302	.384
Total # of Friendly Aircraft Lost: $F(2,14) = 6.637$, $MSe = 2.429$, $p = 0.009$			
Alternate	6.14	5.80	5.20
Current	4.14	6.60	7.80
Composite Score: $F(2,14) = 3.847$, $MSe = 4382.9$, $p = 0.047$			
Alternate	53.8	71.6	124.6
Current	128.0	51.4	51.1

not hypothesize an interface main effect or an interface-by-dimension interaction for weights.

Consistent with our hypothesis, there was a significant main effect for dimensions: $F(5,85) = 4.979$, $MSe = 0.02$, $p < .001$. The main effect for interface was not significant. There was, however, a significant interface by dimension interaction: $F(5,85) = 2.453$, $MSe = 0.004$, $p = 0.04$. As shown in Table 5.7, dimensions MD, TD, and OP received higher weights for the current than alternate system; in contrast, dimensions PD, EF, and FR received higher weights for the alternate system. However, these differences are small and, in our opinion, of minimal importance.

There was also a significant dimension main effect for the (unweighted) ratings: $F(5,85) = 9.797$, $MSe = 608$, $p < .001$. The mean ratings, from highest to lowest workload, were MD (71.11), TD (70.14), EF (68.47), FR (53.19),

TABLE 5.7
Cell Means for Significant ANOVA Effects for TLX

	Weights Only					
	Dimensions					
System	MD	PD	TD	EF	OP	FR
Alternate	.211	.093	.182	.181	.181	.152
Current	.255	.070	.193	.148	.211	.122
Mean	.233	.082	.188	.165	.196	.137

	Weighted Workload Score						
	Dimensions						
System	MD	PD	TD	EF	OP	FR	Mean
Alternate	16.31	6.45	13.61	13.42	9.14	11.91	11.81
Current	18.62	3.81	13.14	9.76	10.69	7.67	10.62
Mean	17.47	5.13	13.38	11.59	9.92	9.79	11.21

OP (51.53), and PD (39.58). In addition, there was an interface main effect: $F(1,17) = 12.809$, $MSe = 316$, $p = .002$. Consistent with the significant interface main effects for response time for aural questions and the percentage of S or V commits, the alternate interface had a higher mean (unweighted) score than the current interface (63.33 vs. 54.67).

The third ANOVA found significant effects for the weighted workload measures too. In particular, there were again significant main effects for dimension [$F(5,85) = 4.907$, $MSe = 124$, $p = .001$] and interface [$F(1,17) = 5.785$, $MSe = 13.29$, $p = 0.028$]. The mean weighted workload ratings, from highest to lowest, were MD (17.47), TD (13.38), EF (11.59), OP (9.92), FR (9.79), and PD (5.13). Averaging across the six dimensions, the alternate interface had a higher mean (weighted) workload score than the current interface (11.81 vs. 10.62). The dimension-by-interface interaction also was significant at the $p < .10$ level: $F(5,85) = 1.984$, $MSe = 34.26$, $p = 0.089$. Examination of the cell means in Table 5.7 shows that the alternate system had a higher weighted workload score for the PD, TD, EF, and FR dimensions and a lower weighted workload score for the MD and OP dimensions.

A fourth ANOVA was run, this time for the 2 (Interface) × 3 (Experience) × 2 (Order) design using the cumulative weighted workload measure as the dependent measure. Although not as strong, there was again a significant Interface main effect: $F(1,12) = 3.52$, $MSe = 62.8$, $p = .085$. The reduction in the strength of the effect was probably due to the partitioning of variance in the design and the reduction in the degrees of freedom for the test. In addition, there was a significant Interface × Order interaction: $F(1,12) = 4.196$, $MSe =$

62.75, p = .063. When the alternate system interface was used first, the cumulative, weighted workload scores for the alternate and current interfaces were about the same (M for alternate = 70.21, M for current = 72.96). However, when the current interface was used first, the cumulative weighted workload score was lower for the current than alternate system (59.06 vs. 71.15).

A fifth ANOVA was run, this time using the 2 (Interface) × 3 (Squadron) design (with 17 participants) and the cumulative weighted workload measure as the dependent measure. The only significant effect was the Interface main effect: $F(1,14)$ = 4.852, MSe = 69.62, p = .045. On the average, there was a higher cumulative weighted workload score with the alternate than current system (70.26 vs. 62.83).

Subjective Workload Dominance (SWORD) Technique. SWORD uses a series of paired comparison questions to assess the relative workload of performing different tasks with different system interfaces. The current study used three tasks (recorrelating symbology, pairing ADFs, and conducting intercepts) and two interfaces (current and alternate). Participants compared the relative workload for each task–interface combination resulting in 15 comparisons. Then, the geometric mean was calculated for each level of each of the two variables. The higher the geometric mean, the higher was the level of subjective workload.

The geometric means were the cell entries to a 2 (Interface) × 3 (Task) ANOVA. Both independent variables were treated as between-subject variables because each participant completed SWORD only once. This first ANOVA included all 18 participants. A second 2 (Interface) × 3 (Task) ANOVA was performed only for participants who passed the consistency test for the paired comparisons. Only 8 of the 18 participants passed the consistency test, suggesting that the paired comparisons were difficult for the participating WDs. The results were comparable and, therefore, are only presented here for the first ANOVA, which used the data for all 18 participants.

First, there was a significant Interface main effect: $F(1,102)$ = 12.498, MSe = 0.008, p = 0.001. Consistent with the previous results, the alternate system interface had a higher mean workload score than the current interface (.197 vs. .136). There also was a significant Task main effect: $F(2,102)$ = 12.128, MSe = .008, p < .001. Pairing ADFs required the least workload (M = 0.132), then recorrelating track symbology (0.141), and the most workload was necessary for conducting intercepts (.227). Third, there was a significant Interface × Task interaction: $F(2,102)$ = 7.626, MSe = .008, p = .001. Examination of Table 5.8 shows that there was less workload with the alternate system for the intercept task and more workload for the ADF and recorrelation tasks.

We also ran an ANOVA for the 2 (Interface) × 3 (Experience) × 2 (Order) design using the overall geometric mean as the dependent measure. The ANOVA was run for 18 participants; we could not run the ANOVA for the 8 participants who passed the consistency threshold because there were not

TABLE 5.8
Cells Means for Significant ANOVA Effects for SWORD

		2 (Interface) × 3 (Task) Design		
		Tasks		
System	ADFs	Intercepts	Recor.	Mean
Alternate	0.168	0.195	0.256	0.206
Current	0.081	0.233	0.063	0.141
Mean	0.132	0.227	0.141	

enough participants given the number of cells in the design. The only significant effect was an Interface main effect: $F(1,12) = 10.882$, $MSe = 0.004$, $p = 0.006$. The alternate system interface had a higher workload level than the current system interface.

We also ran an ANOVA for the 2 (Interface) × 3 (Squadron) design using the overall geometric mean as the dependent measure. We obtained comparable results when we used all 18 participants or when we used only the 8 participants who passed the consistency threshold. In both cases, the only significant effect was an Interface main effect. With all 18 participants, $F(1,14) = 9.712$, $MSe = .005$, $p = .008$; with the 8 participants, $F(1,4) = 9.39$, $MSe = .001$, $p = .037$, M for current = .131, M for alternate = .202.

Subjective Questionnaire. As described previously, a subjective questionnaire was developed to obtain the participants' opinion as to the effect of four features of the alternate system (color, symbols, on-screen menu, and the quasi-automated nomination procedure) on six cognitive processes (attention, workload, memory, situational awareness, situation assessment, and decision making). The questionnaire had one question about each feature-by-process combination, except for the effect of symbols on workload, for which the question was inadvertently left off the questionnaire. A 5-point response scale was used for all questions, with a higher number being more favorable.

A number of ANOVAs were performed in an effort to understand the participants' opinion as to the effect of system features on cognitive processes, as well as the effect of experience on these opinions. In all cases, a between-subjects design was used because the participants completed the questionnaire only once. Although it is not typically done, it may have been more appropriate to have used a within-subject design for our questionnaire data because each participant answered all the questions that comprised the cells of the design.

A 4 (Features) × 5 (Process) ANOVA was run; workload was dropped because of the missing cell. The only significant effect was a Feature main effect: $F(3,328) = 8.9$, $MSe = .605$, $p < .001$. On the average, the symbols were rated most favorably ($M = 3.73$) then color (3.37), on-screen menus (3.16), and, lastly, QAN (3.12).

We then ran a 4 (Features) × 3 (Experience) ANOVA. There was a significant Features main effect: $F(3,381) = 11.282$, $MSe = .666$, $p < .001$. In addition, there was a significant Features × Experience interaction: $F(6,381) = 3.19$, $MSe = .666$, $p = .005$. Table 5.9 presents the cell means for the features by experience interaction. As can be seen, participants in all three experience levels considered symbols and color most favorably. However, the least and most experienced participants considered the QAN feature much more favorably than did participants with a medium level of experience. In contrast, the latter group considered the on-screen menu more favorably than did the least and most experienced participants.

Third, we ran a 3 (Experience) × 6 (Process) ANOVA, including workload. The only significant effect was the Process main effect: $F(5,373) = 3.154$, $MSe = .735$, $p = .008$. The highest mean subjective rating was for decision making ($M = 3.53$), then situation awareness (3.46), attention (3.37), memory (3.33), situation assessment (3.25) and, lastly, workload (2.93). The low subjective rating for the alternate system's effect on workload is consistent with the results for the objective process measures and the two subjective workload measures.

The last ANOVA focused on the participants' ratings for four additional questions. These questions represented hypotheses regarding the effect of each feature on a particular activity performed by the WDs. These four hypotheses were as follows:

- Color would improve WDs' ability to identify the boundaries and geographic features of the map on the AWACS screen.
- The revised symbols would improve WDs' ability to locate high-value targets.
- The on-screen menus would improve WDs' ability to execute critical functions (e.g., commit switch actions and locate the SIF).

TABLE 5.9
Cell Means for the Features × Experience Interaction
for the Subjective Questionnaire

| Experience | Features of the Alternate Interface | | | |
	Color	Symbols	Menu	QAN
Low	3.35	3.63	3.08	3.38
Medium	3.30	3.71	3.34	2.66
High	3.48	3.88	3.00	3.36

- The QAN would produce credible solutions.

The only significant effect for the 4 (Hypotheses) × 3 (Experience) ANOVA was a Hypothesis main effect: $F(3,57) = 16.48$, $MSe = .787$, $p < .001$. The mean values for hypothesis regarding the effect of color was 4.42 (on a 5-point scale); it was 4.63 for the symbol hypothesis, 3.17 for the on-screen menu, and 2.38 for the QAN. Consistent with the other results presented earlier, the on-screen menu and particularly the QAN needed considerably more work.

Interpretation

The analysis revealed a number of important findings. First and foremost, we learned that a mix of interface "features"—the use of quasiautomated displays, the use of selected colors, and the use of alternative symbology, for example—is likely to enhance performance. But we also learned that the strength of the relationships among features and performance is variable and that a whole series of intervening variables—such as training—and other variable feature combinations may explain additional performance variability. The preliminary findings are thus suggestive of all kinds of additional hypotheses and analytical opportunities. Case studies of this magnitude also require lots of resources and relatively complex experimental designs; the work reported here describes a very promising approach to evaluation—and performance enhancement via user-computer interface features design, prototyping and evaluation. Just as importantly, this kind of research may well result in dramatically improved weapons direction performance via a systematic approach to user-computer interface and interaction feature design and prototyping via the synthesis with emerging and advanced information technology.

ACKNOWLEDGMENTS

We would like to thank the US Air Force, Lt. Col. Mike McFarren and Marie Gomes for their administrative and technical support, and Mat Darlrymple and Phil Tessier for their technical contributions to the project. We would also like to thank Dave Klinger and Gary Klein of Klein Associates, Inc. for major contributions to the requirements analysis, prototyping and testing phases of the project.

Real-Time Expert System Interfaces, Cognitive Processes, and Task Performance

Leonard Adelman
Marvin S. Cohen
Terry A. Bresnick
James O. Chinnis
Kathryn B. Laskey

This chapter reports on an experiment that investigated the effect of different real-time expert system interfaces on operators' cognitive processes and performance. The results supported the principle that a real-time expert system's interface should focus operators' attention to where it is required most. However, following this principle resulted in unanticipated consequences. In particular, it led to inferior performance for less critical, yet important cases requiring operators' attention. For such cases, operators performed better with an interface that let them select where they wanted to focus their attention. Having a rule generation capability improved performance with all interfaces, but less than hypothesized. In all cases, performance with different interfaces and a rule generation capability was explained by their effect on cognitive process measures.

There is an increasing body of empirical research demonstrating that the design of information and decision technology can significantly affect operators' cognitive processes and, in turn, performance. For example, Hoadley (1990) showed that color and various tabular and graphic display formats interact to affect information retrieval speed and, in turn, accuracy. Jarvenpaa (1989) showed that graphic formats can affect the nature of information acquisition and evaluation processes and, in turn, performance. Johnson, Payne, and Bettman (1988) also showed that information displays can affect processing strategies and actually cause preference reversals.

The experiment reported herein adds to this growing body of empirical research by demonstrating that a real-time expert system's interface can significantly affect operators' cognitive processes and, in turn, task performance. The chapter is divided into three principal sections. The first section overviews the experiment, in terms of the task, interfaces, and hypotheses guiding it, and then describes how it was implemented. The second section presents the results; the third section discusses them.

THE TASK, INTERFACES, AND HYPOTHESES

We first overview the task in order to provide the context for considering the proposed interfaces and how they were operationalized. Specifically, the task was representative of that facing an Army air defense officer whose job is to identify incoming aircraft as friend or foe and, subsequently, to engage foes. We, however, consider only the identification task.

The identification task requires the integration of information varying in diagnosticity. Sometimes the information is highly consistent in pointing toward friend or foe; at other times it is conflicting. Most of the information enters the air defense system through radars and sensors, and the diagnosticity of this information can be stored in the system. However, other critical information becomes available through alternative means. The operator must integrate the diagnosticity of this information with that already in the system. Sometimes the task occurs under low workload conditions, such as when there are only a few aircraft on the system's display at a given time. However, we are interested here only in the case when the display is filled with aircraft, that is, extremely high workload.

The basic principle guiding the design of an expert system interface for this task was that operators should consider only cases (i.e., aircraft) requiring their attention. The expert system should (a) identify or "screen" routine cases for which its knowledge base is adequate, and (b) direct operators' attention to cases that have either conflicting information or insufficient information for a firm identification. This interface is referred to as *screening*. The screening interface tested herein had three identification categories: firm identifications (i.e., obvious friends or foes), questionable identifications, and unknowns. This kind of screening is similar to, but not identical with, the interface used in the U.S. Army's Patriot air defense system. Preliminary research (Chinnis, Cohen, & Bresnick, 1985; Cohen, Adelman, Bresnick, Chinnis, & Laskey, 1988) found the screening interface to be more effective than either a totally manual or fully automated system for the identification task under extremely high workload.

Based on this design principle, it was predicted that a screening interface would result in higher performance than an interface that identified all aircraft as friend or foe, and then required the operator to make changes by overriding the system's identification. This second interface is referred to as *override.* The override interface is not simply a strawman, but a viable alternative to the screening interface. The override scheme has the advantage of identifying all aircraft, thereby leaving operators free to assess the situation and focus their attention on the cases they—and not the machine—consider most important. Recent research by Klein (1989) on "recognition primed decision making" indicates that experts are extremely proficient at performing situation assessments tasks. In particular, they are adept at recognizing patterns in the data (e.g., the displayed location of friendly and enemy aircraft) that are consistent or inconsistent with how situations should evolve over time. Moreover, these patterns often can be recognized when data are incomplete or conflicting, characteristics of the air defense task. By identifying all aircraft as friend or foe, the override interface gives operators more time than the screening condition does to look for inconsistent data patterns, gather information about particular cases, and make the required changes in identifications. This capability was predicted to be particularly valuable in the high workload condition studied herein. Therefore, high performance levels also were predicted with the override interface.

An experiment was conducted with Army air defense operators to examine the effect of both interfaces on operators' cognitive processes and task performance. In addition, we examined the effect of a rule generation capability that permitted operators to create rules online for the system to use. This capability permitted operators to incorporate their knowledge directly into the system. Because the system's knowledge base left ample room for improvement, we predicted that the rule generation capability would improve performance for both interfaces. However, we predicted that screening + rule generation would result in the highest level of performance.

THE EXPERIMENT

This section is divided into eight parts, dealing with: (a) the experimental design; (b) the participants and site used in the experiment; (c) the experimental testbed; (d) the manual, override, and screening interfaces; (e) the system's rule generation capability; (f) the aircraft simulation used in the experiment; (g) the experimental procedures; and (h) the dependent measures.

Design

The design was a two interface (screening vs. override) × two-rule generation capability (yes vs. no) within-subject factorial design. However, we added a fifth condition: fully manual (i.e., the operator had to make all identifications) with rule generation capability. Preliminary research (referenced earlier) had demonstrated extremely poor performance in the manual condition without the rule generation capability under high workload conditions. We wanted to assess whether providing the rule generation capability alone would make performance comparable to that achieved with any of the other four conditions. Time constraints did not permit running the manual condition without the rule generation capability.

Participants and Site

Fourteen U.S. Army air defense operators participated in the experiment. All participants were either first or second lieutenants who had completed the Air Defense Officer's Basic Course and who had some experience with either the Patriot or Hawk air defense system. The experiment was conducted at the Army Research Institute Field Office at Ft. Bliss, TX.

Testbed

The testbed hardware was an IBM-AT personal computer with a color display monitor. The input device was a three-key mouse. The testbed had a 20-megabyte hard disk for storing the air defense simulation and all software for presenting the experimental materials for each condition and for collecting participants' responses.

Participants were told that they were the tactical control officer (TCO) of a proposed air defense system, and that they had to decide which of the aircraft approaching them were friends or foes. Figure 6.1 shows what the radar screen looked like. Conceptually, the participant, as the air defense operator, was located at the bottom of the screen where the two straight lines come together. The pie-slice-shaped area was the participant's area of responsibility. The participant was protecting two friendly assets.

Aircraft could fly at different speeds and altitudes depending on whether they were bombers, fighters, or helicopters. Aircraft could appear at the extreme top or sides of the screen or they could "pop up" within the sector if they were flying at altitudes below radar detection. If they popped up, they could do so "close" to the participant or along the forward edge of the battle area (FEBA). Two safe-passage corridors were designated for the movement of friendly aircraft, one on the left and one on the right. Aircraft could be outside or inside one of these corridors. If inside a corridor, aircraft could or could not be flying within the altitude and speed parameters set for the corridor. In all cases, the aircraft moved either toward the participant or the sides of the sector, never toward its top.

Participants were told that in order to perform well, they had to correctly identify as friend or foe as many aircraft as possible before the aircraft went off the radar screen. Aircraft went off the screen when they had reached

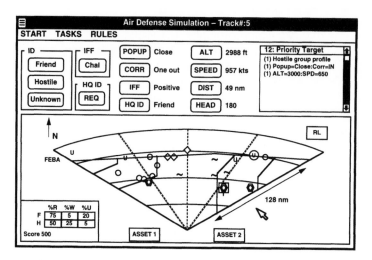

FIG. 6.1. Radar Screen. The test bed's principal display. A "U" represents an unknown, a circle represents a friend, a diamond represents a foe, a zigzag represents a jammer, a square outline marks a current hooked target, and a hexagon outline marks an engaged target.

the sides of the sector or when they were 40 km from the participant's position, which is the closest range ring in Figure 6.1. Participants were told that they were responsible for identifying all aircraft within the FEBA, even though their engagement capability extended only to the second range ring.

Figure 6.1 shows the symbols for different types of aircraft. A black U represented an unknown; a circle represented a friend. A diamond represented a foe; a hexagon outline represented an aircraft that had been engaged by the system; and jammers were indicated by a zig-zag symbol.

Points were used to motivate participants. Participants received 5 points for each correctly identified aircraft. They received no points if they identified the aircraft incorrectly or if it was left as unknown. It was emphasized that their goal was to maximize their point total. Each minute feedback was presented in a box in the lower left-hand portion of the screen. It told the participants what cumulative proportion of friends and foes they had identified correctly, incorrectly, or left as unknown when those aircraft left the screen.

In order to identify an aircraft or obtain more information about it, participants "hooked" the aircraft by (a) using the mouse to guide a cursor to the aircraft, and (b) pressing the mouse's left-hand button. Only one aircraft could be hooked at a time. A hooked aircraft was represented by a square on the radar screen; its track number appeared at the top of the radar display, as shown in Figure 6.1.

Participants could use the mouse-sensitive buttons on the top-half of the screen to obtain more information about a hooked aircraft. In particular, they could find out whether the aircraft popped up (POPUP button); whether it was in or out of the corridor (CORR button); or its speed (SPEED button), distance (DIST button), altitude (ALT button), and heading (HEAD button).

There are two buttons labeled "IFF CHAL" and "HQ ID REQ" in the upper portion of the main display. Participants could gather new information about an aircraft by pressing these buttons. "IFF CHAL" stands for IFF challenge. This is an electronic interrogation signal to which friendly aircraft can respond automatically (unless their equipment has malfunctioned, they do not have an IFF transponder, or their codes are set improperly). Although typical air defense systems simultaneously challenge multiple aircraft, an IFF challenge could be placed only against hooked aircraft in our testbed so that we could examine participants' information-processing strategies. Participants used the "HQ ID REQ" button to ask higher headquarters (HQ) for its identification of the hooked aircraft. The answer to the participant's request appeared to the right of the button. Figure 6.1 shows all the information for the "hooked aircraft" in the right-hand safe-passage corridor.

The participants were told that, in general, there were more foes than friends, and that some of the information (called *cues*) were better discriminators than others. Participants were shown Table 6.1, which indicates which cues were the strongest indicators for each type of aircraft in the simulation. Although the diagnosticity of the cues in our simulation match those in an actual air defense environment moderately well, distinct differences (e.g., that for "HQ ID FOE" or "Corridor Two Out") did develop in the simulation because of the effort to create "test cases," that is, aircraft that would be difficult to identify. We showed participants the diagnosticity table to minimize the extent to which these differences affected participants' performance.

As can be seen from Table 6.1, getting a positive response to an "IFF Challenge" and a friendly response to "HQ ID Request" were the most diagnostic pieces of information. Two actions were taken to prevent participants from simply hooking all aircraft and pressing the buttons for those two pieces of information. First, penalties were attached to "IFF Challenge" and "HQ ID Request." In particular, a point penalty was attached to the former and a time penalty to the latter.

TABLE 6.1
Cue Diagnosticity

Cues Indicating FRIEND		Cues Indicating FOE		
Value	Strength	Value	Strength	Nondefinitive
IFF Positive	Very Strong	Corridor: OUT	Strong	HQ ID Unknown
HQ ID Friend	Very Strong	HQ ID Foe	Strong	Popup (all locations)
Corridor: IN	Strong	Jammer	Strong	Non-Jammer
Corridor: One	Moderate	IFF No Response	Moderate	Corridor: Two
Parameter Out				Parameters Out

Enemy aircraft that can detect the electronic emissions of an IFF challenge can identify and attack the air defense site. This situation is referred to as *exploitation.* It is probabilistic in the sense that not all IFF challenges result in attacks. To represent this situation, and to ensure that participants would use information in addition to IFF challenges, participants were randomly exploited 10% of the times they issued an IFF Challenge. If exploited, they lost 10 points and were notified in the lower left-hand corner of the display. Requesting an HQ ID takes time to perform in an actual air defense environment. To represent this, it took 4 sec for participants to get a response to their requests. During this time they could get information about the hooked aircraft, but they could not hook another aircraft. If the participants performed an IFF CHAL or requested an HQ ID for a hooked aircraft, they could recall the information from their database at a later time without incurring a point or time penalty. This was done by rehooking the aircraft and hitting the IFF and HQ buttons located beneath the SPEED button.

The second action taken was to make information available for certain aircraft so that there would be no need to press the IFF CHAL and HQ ID REQ buttons. This was accomplished by sending participants messages from headquarters. This information appeared in the MESSAGE box at the top, right-hand portion of the display. For example, the message shown in Figure 6.1 indicates that there is a group of hostile aircraft that has popped up close and inside of the safe-passage corridors at an altitude of approximately 3,000 ft (i.e., 914.4 m). No friendly aircraft were at the altitude and speed specified in the messages. Messages were in effect for no more than 10 min, depending on how long it took the group of aircraft to leave the radar screen. After that time, however, other aircraft (e.g., friendly) might appear at the altitudes and speeds identified in a previous message.

A secondary task was used to obtain an objective measure of workload. Specifically, while participants were performing the aircraft identification task, they had to acknowledge orders from higher headquarters. These orders were represented by a light in the upper right-hand corner of the radar display, labeled "RL" in Figure 6.1. This light went on throughout the session based on a Poisson distribution and stayed on for 3 sec. In an effort to ensure that participants responded to the secondary task, (a) participants were told to acknowledge the RL as fast as possible, and (b) lost 2 points if they failed to respond within 3 sec. We obtained accuracy measures by asking participants to press the middle button on the mouse when the light was red and the right-hand button on the mouse when the light was green. The buttons were color coded. Participants obtained 1 point every time they responded correctly within the time limit; they lost 1 point if they responded incorrectly. The testbed was configured so that the mouse could be moved to the system's left-hand side for left-handed participants.

The participants' total "running" score appeared at the bottom of the feedback box in the lower left-hand corner of the display. The total score was a function of (a) the number of aircraft identified correctly, (b) the number of points lost through the exploitation of IFF challenges, and (c) the number of points gained or lost through acknowledgments of the response light. Participants were told that the worst possible score they could obtain at the end of an attack phase was −400 points and that the best possible score was +1,500 points.

Manual, Override, and Screening Interfaces

With the manual interface, participants were told that the system would keep track of all information about the aircraft, but that they had to perform all aircraft identifications. The concept of "conflicting information" was presented in the form of the example of an aircraft that was jamming—which suggests it is a foe—but giving a positive IFF response—which suggests it is a friend. Participants were reminded about how to obtain information through IFF CHAL and HQ ID REQ and about the point and time penalties associated with them.

With the override interface, participants were told that in addition to keeping track of all the information about the aircraft, the system would also make an initial identification of all aircraft based on whether (a) it popped up, (b) it was in the corridor, (c) its speed and altitude met the corridor parameters if it was in the corridor, and (d) it was a jammer. Aircraft initially identified as friends were represented as black circles. Aircraft initially identified as foes were represented as black diamonds. All jammers were initially identified as foes.

The system represented diagnostic information from each cue by a Dempster–Shafer belief function. System identifications were made by combining the belief functions for the observed cues using Dempster's rule to assign beliefs to "friend" and "foe" and classifying the aircraft according to these beliefs.

Dempster–Shafer theory was used because of the ability to represent lack of evidence and conflict. Belief in "friend" and belief in "foe" are not required to sum to unity. The degree to which the sum falls short of 1.0 represents the amount of inconclusiveness or the inability to distinguish between friend and foe. We also made use of a natural measure of conflict, the amount of weight assigned to the empty set before the normalization step in Dempster's Rule. (The reader interested in discussions of Dempster–Shafer theory and comparisons to other formalisms for reasoning under uncertainty is referred to Shafer & Pearl, 1990.)

It is important to reiterate that the system did not have access to messages from headquarters or the results of an HQ ID or IFF challenge when it made the initial identification. Responses to the IFF challenge were, however, included in the system's identification algorithm in an effort to help maintain task representativeness with an actual air defense task. Maximum performance, therefore, required a combination of information available to the machine and information available only to the human.

With the override interface, participants were told about "conflicting information" and reminded about penalties, just as in the manual condition. Participants also were told that unless they changed the system's identification, it would represent their identification when the aircraft went off the screen or if the aircraft was within the FEBA, when the attack session ended. Changes in the identifications made by the participants were color coded. A blue circle represented aircraft that participants identified as friend; a red diamond represented aircraft that participants identified as foe; and a green U represented unknowns.

With the screening interface, participants were again told that the system kept track of all the information about all the aircraft, and that it also made an initial identification of each aircraft based on whether (a) it popped up, (b) it was in the corridor, (c) its speed and altitude met the corridor parameters if it was in the corridor, and (d) it was a jammer. Again, it was noted that the system did not have access to messages from headquarters or an HQ ID, and that it could not initiate IFF challenges.

The system used a blue circle to identify aircraft that clearly appeared to be friends; it used a red diamond to identify aircraft that clearly appeared to be foes. "Firm identifications" were aircraft that had degrees of belief > 0.80 for either friend or foe. The system used a black or purple circle to identify "questionable friends" and a black or purple diamond to identify "questionable foes" with the screening interface. By *questionables* we meant there was not enough information to make a firm identification, but that the evidence was more in favor of one type of identification or another (0.6 < degree of belief < 0.8 for either friend or foe). The color black meant there was no conflicting evidence, just not enough strong data for a firm identification. When there was conflicting information, the system used the color purple. Aircraft with an initial degree of belief < 0.6 for both friend and foe were classified as unknowns (a black U) with the screening interface. This occurred either because of the lack of critical information or because the initial information significantly conflicted.

In the screening condition the system often indicated a "highest priority unknown." The priority rating was based on the amount of uncertainty, the amount of conflict, and the aircraft's "time to nearest friendly asset." This unknown had a solid, purple circle around it. In addition, its identification number appeared at the top of the message box, as shown in Figure 6.1. Participants could hook the "highest priority unknown" by either (a) clicking on its identification number in the message box, which hooked it automatically, or (b) hooking it just like any other aircraft.

Finally, the screening interface had all of the other capabilities available with the override interface. This included making sure that no aircraft left the screen as "unknown." If the participant did not have time to identify the aircraft, the system classified it as friend or foe, depending on which hypothesis had the highest degree of belief.

Rule Generation Capability

The rule generation capability worked the same for the manual, override, and screening interfaces. The system started off with the display shown in Figure 6.2. Assume, for example, that the participant wanted to say that all jammers were to be identified as foes. The participant would move the mouse to the JAM button and click the left-hand button on the mouse. When YES came up, indicating that the participant was referring to jammers, the participant would then go over to the RESULT column in the far left-hand corner of the display and click on "Foe," implying that the participant wanted all jammers to be identified as foes. Then, the participant would click on "Save Rule" to save the identification rule. The system now understands that all jammers are to be identified automatically as foes.

If participants now clicked on RULES at the top of the display, making sure to hold down the mouse button, they would see that a rule called R6(H) had been created. At any time, the participants could move the mouse over R6(H) and lift their finger off the mouse button. The values of the R6(H) identification rule would then appear on the screen, that is, "jammers are foes." If participants wanted to erase this rule, they could do so by clicking on "Clear Rule."

Before creating a new rule, participants selected a rule number. To do this, they (a) clicked on RULES, and (b) while pressing the mouse button,

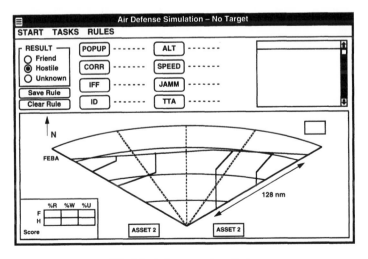

FIG. 6.2. Rule generation display.

moved the mouse to an empty rule and released the mouse button. Initially, rule R6 was selected for the participant because the first five rules in the RULES box were saved for message rules. Message rules were executed first by the system because they refer to rules that were created to identify groups of aircraft with the characteristics indicated in messages. For example, assume participants received the message below:

(1) Foe Group Profile
(1) Popup=No: Corr=TwoOut
(1) Alt=11,000: Speed=1200.

They could create a rule that identified all these aircraft as hostile by doing the following:

(1) Selecting Rule MR1;
(2) clicking on POPUP until it read No;
(3) clicking on CORR until it read TwoOut;
(4) clicking on ALT until it read 10000–80000;
(5) clicking on SPEED until it read 1000+;
(6) clicking on Foe; and
(7) clicking on Save Rule.

This rule would be stored as MR1(H) in the Message category under the RULES menu. It would identify all aircraft as shown with these characteristics. Participants needed to erase or clear this rule from the RULES menu when this message disappeared from the message box because all the hostile aircraft with these (message) characteristics had left the screen.

Participants could create as many rules as they liked before and during an attack phase. The message rules always were executed first. The other rules were executed in the order in which they were listed. It was emphasized that this order was important. For example, if participants wanted all jammers to be hostile and any other aircraft that were traveling correctly within a corridor to be friendly, then the rule for jammers had to come before the corridor rule.

In manual, override, and screening conditions, any aircraft not covered by a rule was handled according to the procedures discussed earlier for the appropriate condition.

Simulation

A 355-aircraft simulation was developed to test the effectiveness of the five experimental conditions. Twenty-five of the 355 aircraft represented leading aircraft, 200 aircraft represented the test cases, and the rest were trailing aircraft. The 25 leading aircraft were used to get all participants actively involved in the task before the test cases appeared. The trailing aircraft ensured that all 200 test cases traveled off the radar scope before the simulation ended. The order of the leading aircraft, test cases, and trailing aircraft was randomized to create three order files so that participants could not remember information about specific aircraft over the course of the day-long session. Then, subjects and conditions were randomly assigned order files. All analyses were performed only for the 200 aircraft representing the test cases; consequently, only the characteristics for these aircraft are considered here.

The 200 test cases were actually composed of three classes (or sets) of aircraft. The classification depended on the degree of belief for the aircraft prior to IFF challenges, HQ ID requests, or messages. The three sets of aircraft directly match the three identification categories in the screening condition: "firm identification" (degree of belief ≥ 0.8), "questionables" ($0.6 <$ degree of belief < 0.8 regardless of conflict), and "unknowns" (degree of belief ≤ 0.6). In particular, there were 85 "firm identification" aircraft, 57 "questionable" aircraft, and 58 "unknowns." In the completely automated condition, the computer correctly identified 87% of the "firmly identified aircraft," 64% of the "questionables," and 46% of the "unknowns."

Development of the simulation required four principal activities. The first activity was to develop these three sets of aircraft. Two domain experts were interviewed to obtain degree-of-belief (DB) values indicating the extent to which each level of each identification cue indicated friend, foe, or was inconclusive. The domain experts were independently interviewed over a number of sessions. Meetings were then held with both domain experts to resolve areas of disagreement. The belief values used to construct the simulation, and used in the override and screening conditions, are shown in Table 6.2. These belief values, and the combinatoric Dempster–Shafer algorithm, were used to create cases for the three identification categories.

The second activity was to determine the values for IFF challenge and HQ ID for aircraft with the same levels on the pop-up, jamming, and corridor cues so that the proportion of friends and foes in the simulation after IFF challenge and HQ ID would approximate the degrees of belief for friend and foe prior to this information. This was important in an effort to ensure the overall representativeness of the simulation, as represented by the system's degree-of-belief values, and yet to create aircraft with conflicting information for the experiment. For example, according to the degree-of-belief values shown in Table 6.2 and the Dempster–Shafer algorithm for combining these values into overall values, an aircraft that, prior to an IFF response or HQ ID, (a) pops up close in the safe-passage corridor, (b) is not jamming, and (c) is at the correct speed and altitude should, in general, has an overall degree of belief (Friend) = 0.64, DB (Foe) = 0.29, and DB (Don't Know) = 0.07. We combined the belief for Foe and Don't Know into a single probability for Foe. Thus, 64% of the aircraft with this initial information were

TABLE 6.2
Degree of Belief Values for Individual Cues Levels
(Assuming Cue Independence)

		Degree of Support		
	Cue	For "Friend"	For "Foe"	For "Don't Know"
IFF	Response	98	0	2
	No Response	0	55	45
JAMMING	Yes	0	95	5
	No	0	0	100
POP-UP	No	0	0	100
	Close	0	80	20
	FEBA	0	60	40
OUT OF CORRIDOR		0	80	20
HEADING IN CORRIDOR	Speed and Altitude Correct	90	0	10
	Speed or Altitude Wrong	75	5	20
	Speed and Altitude Wrong	50	20	30
HQ ID	Friend	95	0	5
	Foe	0	90	10
	Unknown	0	0	100

friends, and 36% were foes. We constructed the responses to IFF challenge and HQ IDs for the 14 aircraft in the simulation with the initial information in the earlier example so that nine of them (i.e., 64%) were actually friends and five of them (i.e., 36%) were foes. In this way, the proportion of friends and foes in the simulation approximated the degrees of belief calculated in the override and screening conditions for the initial information.

The third activity involved in developing the simulation was to create messages so that 28 of the 58 "unknown" aircraft and 28 of the 57 "questionable" aircraft could be perfectly identified without IFF challenges or requests for HQ ID. This was accomplished by having the messages focus on groups of aircraft that were in the safe-passage corridor, but that had particular values for pop-up, speed, and altitude. This action was taken in an attempt to make participants use these identification cues and not perform the task solely on the basis of visual information (i.e., being out of the corridor and/or jamming) or HQ ID and IFF challenges.

The fourth activity was to develop the file of instructions for telling each aircraft where to appear on the screen and how to traverse its path across it. A file of information, referred to as the master file, was created. The master file contained an aircraft's identification number, its cue value for IFF, HQ ID, Pop-Up, Jamming, and Corridor, and its true identification. The master file also contained the aircraft type (fighter, bomber, or helicopter), the aircraft's initial location in terms of distance from the air defense unit and azimuth, its initial heading in degrees north, and its speed and flight path.

Procedures

Each of the 14 participants took an entire workday (approximately 8 hours not including a 1-hour lunch break) to perform the air defense task for each of the five conditions. Most of the time, two participants participated each day the experiment was conducted. The participants were separated by partitions and worked independently on identical testbed systems. In addition, there were two experimenters, one for each participant.

Each participant began the session by reading a job description for the air defense task as represented in the testbed. After they finished reading the description, the experimenter discussed it with them to make sure they understood it, particularly the differences between the testbed and an actual air defense system.

The order with which the participants used the different conditions was counterbalanced. After participants said that they fully understood the job description, they were given the written description for their first condition. After reading it, the experimenter called up a training run and demonstrated it. Then, the participants had an opportunity to work with the condition until they said they felt comfortable with it. The first familiarization session took approximately 30 min because participants had to take time familiarizing themselves with the characteristics of the testbed, particularly using the mouse for all interactions. Also, the first training session using the rule generation capability took about 30 min. Subsequent familiarization sessions for the other conditions took approximately 15 min.

After each familiarization session, participants had a practice session simulating an actual test session. Aircraft appeared on the screen every 4 sec, just as in the test sessions. Participants were urged to perform as well as they could, but not to hesitate in asking the experimenter questions about the system's capabilities. The experimenters observed participants' performance during the practice session, pointing out only those aspects of the written descriptions that the participants were not considering when performing the task. The practice session for each condition took approximately one-half hour.

After answering questions, the experimenter started the test. After completing the test condition, participants completed a questionnaire asking them to evaluate their performance, the workload level, and the degree to which they liked working with the system. After completing the questionnaire, the experimenter would start the next condition. The sequence of "read system description, familiarization session, practice session, test session, and questionnaire" was used for all five conditions.

Dependent Measures

There were four sets of dependent measures in the experiment: objective performance measures; information-processing measures; objective workload measures; and subjective performance, workload, and preference measures. Each set is considered here respectively.

Measurement of participants' performance was straightforward because we knew the true identity of each aircraft. Performance was examined for (a) all 200 test cases, (b) the 85 aircraft with a degree of belief ≥ .80, (c) the 57 aircraft with a degree of belief between .60 and .80, and (d) the 58 aircraft with a degree of belief ≤ 0.60 before participants collected additional information via IFF, HQ ID, messages, or other available sources. Groups (b), (c), and (d) represent, respectively, the "firm identifications," "questionable identifications," and "unknowns" classifications used in the screening interface.

There were two cognitive process measures: (a) the percentage of aircraft examined (i.e., "hooked") by the participants, and (b) the average length of time (in sec) for each hooking. These measures were used to infer how the interfaces and rule generation capability affected operators' decisions to examine specific cases and the amount of time they took to process information about them, respectively. Both measures were obtained by capturing participants' mouse clicks while they performed the task.

The two objective workload measures depended on the secondary task performed by the participants. The first workload measure was the participants' response time to the "response light." The second measure was their response accuracy; participants were told to press the middle button of the mouse when the light was red and the right-hand button of the mouse when the light was green.

Two questionnaires were used to obtain participants' subjective performance, workload, and preference measures. One questionnaire was given after the completion of each condition and asked participants to rate (on a 9-point scale) their performance, the level of workload, and the degree to which they liked working with the system. The second questionnaire was given at the end of the experimental session, immediately after the participants completed the questionnaire for the last condition. It generated results comparable to the first questionnaire and is not considered here.

The results are now presented for each of the four sets of dependent measures. Within-subject Analyses of Variance were performed throughout.

RESULTS

Performance

Table 6.3 presents the performance data for each of the five experimental conditions. Performance was defined as the mean percentage of correct identifications. It is presented for (a) all 200 test cases, (b) "firm identifications" (initial degree of belief ≥ 0.80), (c) "questionable identifications" (.60

TABLE 6.3
Performance in Terms of Mean Percent Correct,
as a Function of Experimental Condition

		All 200 Aircraft	Firm IDs (DB ≥ .8)	Quest. IDs (0.6 < DB < .8)	Unknowns (DB ≤ .6)
No Rule	Screening	77.4	87.2	67.2	74.2
Generation	Override	75.1	88.7	72.1	59.6
Capability	Manual[a]	33.0	36.1	36.7	24.5
Rule	Screening	80.9	88.6	76.1	75.6
Generation	Override	80.0	89.7	77.1	69.7
Capability	Manual	68.6	78.7	66.2	57.4
Totally Automated		69.0	87.0	64.3	46.6

[a]Performance data from an earlier experiment.

< initial degree of belief < 0.80), and (d) "unknowns" (initial degree of belief ≤ .60). In addition, Table 6.3 presents the performance data for two other conditions. The first condition is the manual condition without the rule generation capability. The data for this condition come from an earlier experiment (Cohen et al., 1988). The second condition is the totally automated system, that is, without the operator, and is based solely on the system's knowledge base. Both conditions are for comparison purposes only; they were not included in any statistical analyses.

Examination of Table 6.3 led to three conclusions. First, as predicted, when there was no rule generation capability, the screening interface led to higher overall performance than did the override interface [$F(1,13) = 5.68$, $p = .03$, $SSe = .017$]. However, the performance enhancement was small, only 2.3%. Moreover, it was not uniform over the three identification categories. In particular, screening without rule generation resulted in a large performance improvement for the "unknowns" [$F(1,13) = 40.56$, $p < .001$, $SSe = .098$]. In contrast, it was significantly worse than override without rule generation for both the "questionable identifications" [$F(1,13) = 5.97$, $p = .03$, $SSe = .153$] and the "firm identifications" [$F(1,13) = 7.82$, $p = .02$, $SSe = .005$].

Second, adding a rule generation capability increased performance for all three interfaces over all identification categories. However, the only increase that was statistically significant for the screening interface was for the "questionable identifications" [$F(1,13) = 6.07$, $p = .03$, $SSe = .237$]. And the only increase that was statistically significant for the override interface was for the "unknowns" [$F(1,13) = 19.16$, $p = .001$, $SSe = .102$]. These results indicate that the operators were able to use the rule generation capability to improve their performance for some of the types of aircraft for which they were previously having trouble with each interface. Contrary to our predictions, there was no statistical difference in total level of performance between the screening and override interfaces when the rule generation capability was available.

Third, adding a rule generation capability to the manual interface resulted in a stunning increase in performance with that interface, as compared with results from an earlier experiment (Cohen et al., 1988) that included that condition. However, the total performance level was still significantly lower than that for override without rule generation [$F(1,13) = 4.44$, $p = .055$, $SSe = .177$], the lowest other condition. In addition, it resulted in performance levels at least 10% lower for each identification category than those achieved by adding the rule generation capability to the screening and override interfaces. Finally, the manual interface with rule generation capability did not, in total, outperform the totally automated condition. Although it performed better for "unknowns," it performed worse than the totally automated condition for "firm identifications."

These results suggest that, at least for the current task, operators needed both the preexisting knowledge base (i.e., automated identification) and a rule generation capability to achieve their highest levels of performance. Even so, operators were not able to achieve the maximum level of performance with any of the conditions. The simulation was constructed so that 97.5% of the test cases could be correctly identified on the basis of all the information. In short, there was still room (approximately 15%) for improvement.

Cognitive Processes

Table 6.4 presents the mean percentage of aircraft hooked in each of the five experimental conditions for all 200 test cases and for each of the three identification categories. Data are not presented for (a) the manual without rule generation capability because it was not available, nor (b) the totally automated condition because there was no operator interaction.

Examination of Table 6.4 led to two conclusions. First, the interface condition affected the type of aircraft hooked. On the average, operators hooked

TABLE 6.4
Mean Percentage of Aircraft Hooked

		All 200 Aircraft	Firm IDs (DB ≥ .8)	Quest. IDs (0.6 < DB < .8)	Unknowns (DB ≤ .6)
No Rule	Screening	45.1	10.8	42.6	82.0
Generation	Override	52.5	29.2	72.9	55.4
Capability					
Rule	Screening	33.2	8.7	32.4	58.4
Generation	Override	39.9	26.3	46.9	46.4
Capability	Manual	41.1	29.2	42.6	51.4

significantly more "unknowns" with the screening than the override interface [$F(1,13) = 25.25$, $p < .001$, $SSe = 1.06$]. In contrast, they hooked significantly more "questionable identifications" [$F(1,13) = 17.89$, $p < .001$, $SSe = 2.32$] and firm identifications [$F(1,13) = 29.27$, $p < .001$, $SSe = 0.81$] with the override than screening interface. This effect was particularly pronounced when there was no rule generation capability and is perfectly consistent with the performance data presented in Table 6.3.

Second, having the rule generation capability reduced the percentage of aircraft hooked with both the screening and override interfaces. This was particularly pronounced for the "unknowns" [$F(1,13) = 15.21$, $p = .002$, $SSe = 1.29$] and the "questionable identifications" [$F(1,13) = 25.41$, $p < .001$, $SSe = 1.07$]. The biggest reduction in hooking "unknowns" occurred for the screening interface. In contrast, the biggest reduction for "questionable identifications" was for the override interface. These results suggest that operators tried to reduce their workload by generating rules that would focus the system's attention to those objects they had previously focused on most.

Table 6.5 presents the mean length of time (in sec) per aircraft hooking. We make two observations. First, operators took longer to examine aircraft with the screening than override interface for all identification classes. This increase was significant at the $p < .05$ level for "unknowns" and "questionable identifications," but not for the "firm identifications." Second, adding the rule generation capability to the screening and override interfaces increased the examination time per hooking for all identifications categories, but the increase was not significant ($p = .10$) for the "firm identifications." With the screening interface, the increase in the examination time per aircraft was most pronounced for "firm identifications" and "questionable identifications." With the override condition, the increase was most pronounced for the "unknowns."

Again, these cognitive process findings track with the performance data. The rule generation capability permitted operators to (a) have the system focus on those objects that they were previously focusing on and, thereby (b) have more time to focus their attention on where they thought it was

TABLE 6.5
Mean Length of Time (in Sec) per Hooking

		All 200 Aircraft	Firm IDs (DB ≥ .8)	Quest. IDs (0.6 < DB < .8)	Unknowns (DB ≤ .6)
No Rule	Screening	11.18	12.40	10.25	10.90
Generation	Override	9.34	8.97	9.46	9.58
Capability					
Rule	Screening	16.83	20.63	15.37	14.49
Generation	Override	11.67	10.18	11.97	12.84
Capability	Manual	13.40	12.92	13.49	12.92

required. Except for "firm identifications," this permitted the operators to significantly improve performance for the identification category giving them trouble with each type of interface.

Adding the rule generation capability to the manual interface resulted in scores for both cognitive process measures that were in between those obtained with the screening and override interfaces with this capability, for each identification category. These findings again indicate that the type of interface affected operators' cognitive processes. However, the performance data indicate that performance was not solely a function of cognitive processes. One needed the content provided by the rules contained in the system, as well as those generated by the operator.

Workload

Table 6.6 presents the means for all five conditions for both objective workload measures. The only significant paired comparisons were greater response accuracy [$F(1,13) = 4.27$, $p = .059$, $SSe = 835.43$] and faster response time [$F(1,13) = 11.65$, $p = .005$, $SSe = 330341.4$] for the screening interface without the rule generation capability than with it. These results are consistent with the large increase in examination time per hooking when the rule generation capability was added to the screening interface.

Subjective Measures

Table 6.7 presents the mean subjective performance, workload, and preference responses for each condition. Higher values represent better scores on all three measures.

The rule generation capability had a positive effect on the mean subjective performance ratings for all three interfaces, a result that agrees completely with the objective performance data. A one-way, repeated-measures ANOVA on the subjective performance ratings for the five conditions was significant [$F(4,52) = 4.35$, $p = .004$, $SSe = 80.95$]. Paired comparisons showed that mean

TABLE 6.6
Mean Values for Both Objective Workload Measures for the Five Interfaces

	Mean Accuracy of Response to "Response Light"	Mean Speed (in msec) of Response to "Response Light"
Override: No Rule Generation	89.21	1196.6
Screening: No Rule Generation	89.07	1179.1
Manual: Rule Generation	85.71	1283.9
Override: Rule Generation	87.57	1241.9
Screening: Rule Generation	84.64	1324.6

TABLE 6.7
Mean Subjective Performance, Workload, and Preferences

		Rule Generation Capability	
		No	Yes
(A) Subjective Performance			
Human-Machine	Manual	3.73[a]	5.36
Interface	Override	5.50	6.93
Factor	Screening	5.71	6.57
(B) Subjective Workload			
Human-Machine	Manual	3.47[a]	4.14
Interface	Override	4.50	6.07
Factor	Screening	5.57	5.43
(C) Subjective Preference			
Human-Machine	Manual	4.67[a]	4.79
Interface	Override	4.50	6.64
Factor	Screening	5.20	6.14

[a]Data from high workload: manual condition in previous study.

subjective performance for "override + rule generation" was not significantly different from "screening + rule generation," but both were significantly higher than all other conditions. No other paired comparisons were significant.

In addition to having the best mean subjective performance rating, "override + rule generation" had the best mean subjective workload rating. In contrast with the objective workload measures, it was significantly better than that obtained for both the manual without rule generation [$F(1,13) = 8.37$, $p = .013$, $SSe = 80.93$] and override without rule generation [$F(1,13) = 13.44$, $p = .003$, $SSe = 33.43$] conditions. However, it was not significantly better than that obtained for either screening condition. And, in contrast with the objective workload measures, adding the rule generation capability to the screening interface did not increase subjective workload.

Examination of the preference data again shows the positive effect of the rule generation capability, although the effect for the manual interface is small. The mean subjective preference rating for "override + rule generation" was significantly higher than that for all conditions except "screening + rule generation" at the $p < .05$ level. These results are identical to those obtained for the subjective performance rating.

Interpretation

The experiment reported herein demonstrated that the type of interface for a real-time expert system can significantly affect task performance. As predicted, the screening interface led to significantly better overall performance than the override interface. However, the performance enhancement was small. Moreover, it was not uniform over the three identification categories. For although performance with the screening interface was better for "unknowns," it was worse than override for "questionable identifications" and "firm identifications."

Performance with the two types of interfaces was directly tied to the type of aircraft hooked, a cognitive process measure. Operators hooked significantly more "unknowns" with the screening than override interface. In contrast, they hooked significantly more "questionable identifications" and "firm identifications" with the override interface. These results indicate that operators' performance with the interface was mediated by the interface's effect on operators' cognitive processes.

The cognitive process measure also shed light on how adding the rule generation capability to the screening and override interfaces affected performance. Specifically, adding a rule generation capability increased performance for both interfaces for all identification categories. However, the only performance increase that was statistically significant for the screening interface was for the "questionable identifications." And the only increase that was statistically significant for the override interface was for the "unknowns."

These performance data correlate with (a) increases in the examination time per hooking for "questionable identifications" with the screening interface and "unknowns" with override, and (b) decreases in the percentage of aircraft hooked in the other categories—specifically, "unknowns" with screening and "questionable identifications" with override—with a rule generation capability. Taken together, the two cognitive process measures suggest that the rule generation capability permitted operators to make the system focus its (new) knowledge and attention where they used to, thereby letting them take more time to consider the cases requiring their attention. A significant byproduct of this capability was essentially equivalent and better subjective performance, workload, and preference ratings for the screening and override interfaces when the rule generation capability was available.

Adding a rule generation capability to the manual interface resulted in a large increase in performance with that interface. However, the total performance level was still significantly lower than that achieved in the other four conditions. In addition, the manual interface with rule generation capability did not, in total, outperform the totally automated condition. All other conditions did. These results indicate that operators needed both the knowledge base and rule generation capability to achieve their highest levels of performance.

Contrary to our predictions, "screening + rule generation" did not result in a significantly higher total level of performance than that achieved with "override + rule generation." The examination time data suggest that this occurred because operators spent significantly more time per hooking, with "screening + rule generation" than "override + rule generation" for all identification categories. Our post-hoc hypothesis is that if the time per hooking were reduced, performance would have been significantly better in the "screening + rule generation" condition. There was ample room for improvement, for performance was still 15% lower than the maximum level possible. However, the significantly higher workload for both objective measures with the "screening + rule generation" condition suggests that it might be difficult to improve cognitive processing speed and, in turn, performance.

In closing, we make five general points. First, task characteristics, in terms of the percentage of aircraft that were "unknowns," "questionable identifications," and "firm identifications," may have had a significant impact on overall performance. For example, without the rule generation capability, screening resulted in significantly better performance for the total 200-aircraft simulation because its large performance advantage for "unknowns" compensated for its smaller, yet significantly poorer performance for both other identification categories. In the current simulation, 42% of the aircraft were "firm identifications," 29% were "questionable identifications," and 29% were "unknowns." However, if the simulation had been 90% "firm identifications" and "questionable identifications," it is possible that the override interface would have resulted in significantly better total performance. This hypothesis requires empirical investigation. However, it suggests that task characteristics need to be considered when evaluating system interfaces.

Second, the study provided support for the cognitive system engineering principles guiding the design of both the screening and override interface. Regarding the former, the interface to a real-time expert system should direct operators' attention to those cases most in need of it. When done in the screening condition, this resulted in improved performance for these cases (i.e., "unknowns"). However, it impaired performance for less critical but still important cases. Therefore, the interface must still give operators the time to assess the situation and select cases they want to consider. Such a balancing act is not an easy one and usually will require an iterative development and evaluation process to implement successfully.

Third, adding a rule generation capability to the screening interface represents one approach to improving this balancing act. It permitted operators to expand the set of cases handled by the system, thereby giving them more time to select and examine the cases they desired. However, the performance enhancement increased operators' examination time per case and their objective workload. Future applications need to minimize this unanticipated negative consequence to realize the full performance potential of adding a rule generation capability to the screening interface.

Fourth, further examination of performance with an override interface and rule generation capability is also warranted. Operators performed well with this combination. They knew where their attention was needed but, like the "screening + rule generation" condition, increased their examination time per case. Consequently, this condition did not realize its full performance potential either.

Fifth, the field of cognitive engineering is still in its infancy. A growing body of research shows that system design does affect cognitive processing and, in turn, performance. We are beginning to develop general principles that can be used to guide design. However, as the experiment reported herein demonstrates, following these principles can result in unanticipated consequences. Consistent with the position of a growing body of researchers (e.g., Adelman, 1992; Andriole, 1989; Gould & Lewis, 1985; Sage, 1991), experiments must be a critical element of the system development process, simply because we do not know the specific effect of many system features on operators' cognitive processes.

ACKNOWLEDGMENTS

The research was supported by the U.S. Army Research Institute for the Behavioral and Social Sciences, Contract No. MDA903-85-C-0332 to Decision Science Consortium, Inc. (DSC), under the Defense Small Business Innovative Research (SBIR) program. We are grateful to our scientific officers, Irving Alderman and Laurel Allender, for helpful guidance and assistance; to Russell J. Branaghan and Mark J. Sugg, graduate students in the Department of Psychology at New Mexico State University, who were the experimenters for the two experiments at Fort Bliss; and to our colleagues at Decision Science Consortium, Inc., Monica M. Constantine, John Leddo, Jeffrey S. Mandel, F. Freeman Marvin, Lynn Merchant-Geuder, Bryan B. Thompson, and Martin A. Tolcott, for their contributions to the work.

Information Order Effects on Expert Judgment

Leonard Adelman
Terry A. Bresnick
Paul K. Black
F. Freeman Marvin
Steven G. Sak

This chapter reports on some experiments designed to investigate a belief-updating model developed by Einhorn and Hogarth (1987) and Hogarth and Einhorn (1992). The model predicts that when information is presented sequentially and a probability estimate is obtained after each piece of information, people will anchor on the current position and then adjust their belief on the basis of how strongly the new information confirms or disconfirms the current position. Moreover, they hypothesized that the larger the anchor, the greater the impact of the same piece of disconfirming information. Conversely, the smaller the anchor, the greater will be the impact of the same piece of confirming information. This differential weighing of the most recent piece of information is predicted to result in different anchors from which the adjustment process proceeds and, in turn, different final probabilities based on the sequential order of the same confirming and disconfirming information.

Hogarth and Einhorn's experiments were with college students performing tasks for which they were not necessarily experts. However, there is also empirical support for the model with experts performing the tasks for which they were trained. For example, Ashton and Ashton (1988, 1990) presented results consistent with the model's predictions using professional auditors. Serfaty, Entin, and Tenney (1989) presented results showing an order-effect bias for Army officers performing a situation-assessment task. And Adelman, Tolcott, and Bresnick (1993) and Adelman and Bresnick (1992) found results consistent with the model's predictions for Army air defense operators using a paper-and-pencil instrument and air defense simulators, respectively.

Figure 7.1 presents a pictorial representation of the model's predictions for the Army air defense domain, which is the setting for the experiment reported here. In particular, Figure 7.1 illustrates the case in which an air defense operator sequentially receives three different pieces of information about an unknown aircraft.

We assume that after the first piece of information, the operator thinks the probability that the unknown aircraft is a friend is 0.65. For Order 1, the operator then receives confirming information further suggesting that the aircraft is a friend, and then disconfirming information suggesting that it is a foe. For Order 2, the second and third pieces of information are presented in the opposite order. Receiving the same information in different ordered sequences is quite possible, depending on an aircraft's flight path and actions and the way the information is relayed to the operator.

The prediction is that after seeing all pieces of information, operators will give a significantly higher probability of friend to Order 2 (i.e., Disconfirm/Confirm) than Order 1 (Confirm/Disconfirm) because of the differential weights assigned to the information based on the different anchors. In particular, the confirming information will receive a stronger positive weight, and the disconfirming information a weaker negative weight, in the Disconfirm/Confirm rather than the Confirm/Disconfirm sequence. This is represented pictorially in Figure 7.1 by what Einhorn and Hogarth (1987) called a "fishtail." As noted earlier, these predictions were supported in the previous Army air defense studies. Moreover, when the simulator was used, we found that the significantly lower mean probability of the Friend, P(F), estimate for the Confirm/Disconfirm order was accompanied by a significantly larger number of engagements for a restricted set of aircraft. Engagements was not a dependent variable for the paper-and-pencil task.

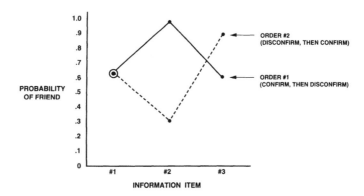

FIG. 7.1. Representation of Hogarth-Einhorn Model predictions.

Simulators are as close as one can get to a fully representative setting for performing controlled research testing the Hogarth–Einhorn model for the Army air defense task. The generality of the results reported by Adelman and Bresnick (1992) are important because task characteristics should not affect how air defense operators process information. Regardless of whether information is presented sequentially or all at once, or whether probability estimates are assessed or not, air defense operators are supposed to process information in the aggregate by considering all confirming and disconfirming information. That is how they are trained, and that is how automated Army air defense algorithms make their recommended identifications. The fact that task characteristics did affect how operators processed information, and that the effects were consistent with the predictions of the Hogarth–Einhorn model, are of practical significance.

The experiment reported herein attempts to replicate and extend the findings reported in Adelman and Bresnick (1992). Replication is necessary because not all of the cells in the previous experiment were implemented perfectly on the air defense simulator, for it was not designed for controlled experimentation. Moreover, the order effect was not found for all cases, just for aircraft whose initial information suggested they were friendly. In addition, the order effect was mediated by the operators' experience. Patriot air defense officers with European experience did not show the effect; officers without such experience did. In the paper-and-pencil task, officers did not show the effect; enlisted personnel did. Therefore, the generality of the obtained order effect is a research issue.

The experiment also attempted to extend the previous research by testing whether a theoretically derived modification to the operator's display could eliminate the order effect, assuming it occurred. In particular, the goal was to design a display modification to foster a more global-processing strategy instead of an anchoring and adjustment heuristic. According to the Hogarth–Einhorn model, this should be accomplished by simultaneously displaying all information and grouping separately that which supports Friend and that which supports Hostile. This is not routinely done with the Patriot display, although track histories can be requested by the operator. The hypothesis was that by displaying all of the essential information at all times, operators would be less likely to differentially weigh the most recent piece of information on the basis of a previously developed anchor point.

METHOD

Participants

Twenty trained tactical control officers (TCOs) for the Patriot air defense system participated in the experiment. Officers were used because Adelman et al. (1993) found them less susceptible to an order-effect bias than enlisted personnel. Moreover, they were responsible for the identification task.

Eighteen of the 20 participants had participated in Desert Shield/Desert Storm. The group included 1 captain, 8 first lieutenants, and 11 second lieutenants. The average amount of experience with the Patriot system was 1.9 years; only four participants had more than 2 years of experience. No analysis examined participants' characteristics, for we considered the group to be homogeneous in its experience. It should be noted that we attempted to include a second group of less experienced Patriot officers, but were unable to do so because of scheduling problems. In addition, we were able to obtain only two participants with European experience.

Design

Independent Variables. Four independent variables, with two levels on each variable, were used in a fully within-subject design. By crossing the first three independent variables described later, we created the eight different aircraft tracks displayed to the participants. The fourth independent variable was whether or not the participants had access to the theoretically derived display when evaluating the tracks.

The first independent variable was whether the first piece of information about an aircraft indicated that it was friendly or hostile. The second independent variable was the order in which confirming and disconfirming information was presented early in an aircraft's track history. For Early Order 1, the second item of information confirmed the initial information, and the third item of information disconfirmed it (CD). The opposite order was used for Early Order 2, that is, disconfirm then confirm (DC). The third independent variable was the order in which confirming and disconfirming information was presented late in a track's history. For Late Order 1, the fourth item of information confirmed the first information item, and the fifth item was disconfirming (CD). For Late Order 2, the opposite order (DC) was used for the fourth and fifth information items. One of the eight tracks is described in detail in the Procedures section.

The fourth independent variable was whether or not the participants used the theoretically derived display. The display attempted to foster a more global-processing strategy in which all information was considered for all judgments. We attempted to accomplish this in two ways. First, all available information about a track was presented at all times. In addition, the display showed which information was current (*current cues*) and which information was historical (*prior cues*). Second, the display indicated the general diagnosticity (or weight) of each cue based on Patriot's internal algorithm. The Patriot algorithm is basically an additive rule. Several cues are weighed by Patriot as indicative of a friendly aircraft, and the algorithm adds positive points to an aircraft's score depending on the strength of the cue. Other cues indicate a hostile aircraft, and the algorithm adds negative points to the aircraft's score depending on the strength of the cue. The absence of these cues results in no points. An example of the display is presented in the Procedures section.

Dependent Variables. Three dependent variables were used to assess if the independent variables affected the participants' judgments. The first dependent variable was the probability estimate participants gave for whether the aircraft was a friend or a hostile. The second was the participants' actual identification of the aircraft as friend, hostile, or unknown. It was quite possible for information order to significantly affect the probabilities, but for these differences to be small and of minimal practical significance such that there was no significant difference in participants' identifications. The identification, not the probability estimate, is the more important measure of air defense officers' judgments.

The third dependent variable was the participants' decision to engage an aircraft or not. The engagement decision is correlated with, but not identical to the identification judgment. The engagement decision also is affected by other factors, such as an aircraft's perceived threat. However, it is the ultimate behavioral measure of air defense officers' judgments.

Procedure

The experiment was implemented on the Patriot training simulators at Fort Bliss, TX. Each session had two to four participants. The simulation system displayed the aircraft tracks to session participants at the same time. However, participants worked independently at separate Patriot consoles and were free to process the tracks as they thought best.

The session began with an introduction and brief overview of what probability estimates of friend or hostile meant. Patriot officers do not routinely make probability estimates, although as college graduates, most of them are familiar with the probability scale. After the overview, we used a paper-and-pencil instrument in which participants were asked to assess the probability that individual events (such as an aircraft being inside the safe-passage corridor or jamming friendly radar) indicated that the aircraft was friendly or hostile. The estimates were reviewed prior to proceeding to ensure that they were consistent with general knowledge regarding the relative predictability of the events and, therefore, that the participants could make the requested probability estimates for the study. This procedure was used successfully in Adelman and Bresnick (1992).

A southwest Asian scenario was presented to provide a context for the session. However, the display provided no unique geographic information.

The cover story justified why participants were working in an autonomous mode, that is, without communications with headquarters or other units. By being in an autonomous mode, the participants were solely responsible for their identification and engagement decisions.

Participants were also told that there had been a few instances of compromise of the IFF Mode 3 response to Patriot's electronic interrogation signal; consequently, this response was not a perfect indicator of friendly aircraft. The cover story also indicated that friendly and hostile aircraft were equally likely to appear in their sector in order to nullify participants' perceptions of this prior information. Finally, Patriot was operated in the semi-automatic engagement mode. In this mode, Patriot will make identification recommendations automatically based on its interpretation of the same information seen by the officers, but will not automatically engage aircraft it classifies as hostile. This ensured that all data were collected for all participants throughout the flight path for all tracks. It also ensured that all participants knew Patriot's identification and engagement recommendations.

One aircraft track was displayed on the Patriot console at a time. The display and console were stopped after each information item for 15 sec for data collection. This ensured that all data were collected at the same point in a track's history for all participants. The data were collected by an observer working with each participant. At each stopping point, the participant (a) told the observer the probability that the aircraft was friendly (or hostile); (b) identified the aircraft as friend, hostile, or unknown; and (c) indicated whether or not they would engage the aircraft. Participants thought aloud throughout the session. If at any time in a track's history they indicated that they would engage the aircraft, the aircraft was considered engaged. However, they were told not to actually engage the aircraft. Thus, we could continue to collect data throughout the track's flight path, thereby ensuring that we had a complete data set for all participants. In addition, it permitted us to capture cases in which the operators reversed their decision in the face of subsequent information.

Figure 7.2 illustrates how this procedure worked for one of the eight tracks in the design. In particular, this is the track whose (a) initial information suggests it is hostile, (b) early order sequence is Disconfirm/Confirm (D/C), and (c) late order sequence is Confirm/Disconfirm (C/D). The air defense sector is represented by the inverted pyramid. The operator is located at the apex (bottom of figure) and is protecting two assets.

The aircraft first appeared on the Patriot display at point 1 and began jamming friendly radar. After the aircraft crossed the Fire Support Coordination Line (FSCL), the operator was told that the IFF interrogation capability was inoperative; consequently, IFF information was not immediately available because of a "system problem." The Patriot console was frozen, and the first piece of data was collected at point A. After 15 sec, the console

TRACK 1

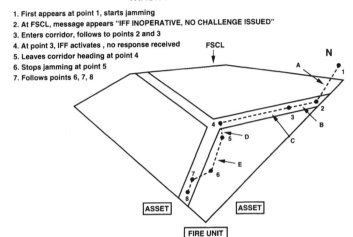

1. First appears at point 1, starts jamming
2. At FSCL, message appears "IFF INOPERATIVE, NO CHALLENGE ISSUED"
3. Enters corridor, follows to points 2 and 3
4. At point 3, IFF activates , no response received
5. Leaves corridor heading at point 4
6. Stops jamming at point 5
7. Follows points 6, 7, 8

FIG. 7.2. Representation of one of the eight aircraft tracks in the design, as depicted on the observers' coding form.

was unfrozen and the aircraft track continued to point 2, where the aircraft entered and began following the safe passage corridor. This is what a friendly aircraft would do and, therefore, represented disconfirming information to the original hypothesis of hostile. The console was frozen and data were collected at point B. After the console was unfrozen, the aircraft proceeded to point 3, where the operator was informed that the IFF was active again and that no response had been received from the aircraft after its interrogation. This information suggests that the aircraft is hostile and confirms the initial piece of information about it. Thus, the early order sequence was Disconfirm/Confirm.

The late order sequence was Confirm/Disconfirm for this track. After the console was unfrozen, the aircraft continued down the safe-passage corridor to point 4, where it left the corridor. Leaving the corridor is a hostile cue and, thus, confirms the initial information. The console was frozen and data collected at point D. After the console was unfrozen, the aircraft stopped jamming friendly radar at point 5. This is a friendly cue and, therefore, disconfirmed the initial information. The console was frozen soon thereafter, and the data collected at point E. After the console was unfrozen, the aircraft proceeded down the remainder of the path shown in Figure 7.2. Data were collected one more time, around point 6, but only the data collected from points A through E were used in the analysis testing for order effects.

Two steps were taken to minimize participants' ability to recognize previous tracks that had the same information presented in different orders. First, six other tracks were interspersed with the eight design tracks. These are referred to later as filler tracks. Second, we created equivalent variations of the tracks used in the design. This was accomplished in a two-step procedure. The first step was to create mirror images of the tracks. For example, there were two tracks that had the first piece of information indicate friend and a Confirm/Disconfirm early order sequence. Because these two tracks had the same first three pieces of information, we had one of the tracks appear on the left-hand side of the Patriot display and the other on the right-hand side; that is, the first half of one track was a mirror image of the other. They had, of course, different late order sequences. Therefore, in the second step, we designed the flight paths so that late order differences were not affected by the aircraft's position on the display. As a result, the tracks looked very different. However, each set of four friends and four hostiles was equivalent except for the order in which the participants received information.

Although they occur infrequently, situations with (a) only one or two tracks on an air defense display, and (b) tracks with conflicting information are representative of real-world situations. When the U.S.S. Vincennes shot down an Iranian airliner, there were only two aircraft tracks on its display, and the information about the airliner's track was conflicting. During the Gulf War, there was often only one track on the Patriot display at a time. Although these tracks represented high-speed tactical ballistic missiles, the identification and engagement decisions were similar to those in the experiment.

As mentioned earlier, the experiment was a fully crossed, within-subject design. Therefore, all the tracks were presented to each participant twice: once with the theoretically derived display, and once without it. Participants were told that they were processing the tracks for two distinct scenarios. The scenarios were the same except for (a) the inclusion of two different filler tracks, and (b) the presentation of the tracks in a different order. Half the participants performed the first scenario in the session using the display; half used the display for the second scenario. In addition, the ordered sets were counterbalanced between the first and second scenarios to further reduce the possibility of presentation order confounding the results. It was not possible to implement the full randomization of all tracks to all participants because of the logistical constraints of the Patriot simulator.

The display was not integrated into the Patriot system, for the software development costs that would have been required were substantially beyond those available. Instead, the display was presented via a portable computer positioned alongside the Patriot display. Information was presented on the display by the observer pressing the arrow keys immediately after the cue was observed on the Patriot display.

Figure 7.3 shows what the added display looked like at the fifth data collection point (i.e., point E) for the track shown in Figure 7.2. The top

TRACK: 1 E	CUES POINT TOWARDS:		
	Friendly		Hostile
CURRENT CUES		LOST SPC STOP JAM	NO IFF
PRIOR CUES	SPC		JAM

CUE DIAGNOSTICITY TABLE

Friendly		Hostile	
▨▨▨ VALID MODE 4 IFF VALID MODE 1, 3 IFF IN SPC	NO SPC NOT JAMMING STOPPED JAMMING	JAMMING RESTRICTED VOLUME NO IFF RESPONSE	▨▨▨

FIG. 7.3. Theoretically derived display.

half of the display presents the current and prior information and indicates whether the information points toward the aircraft being friendly or hostile. In particular, the display shows that the aircraft (a) was not currently in the safe-passage corridor (SPC), although it had been; (b) had stopped jamming; and (c) had not responded to the IFF interrogation. The latter cue, and jamming, are indicative of a hostile aircraft.

Examination of the bottom half of the display shows the relative strength of the cues. The two friendly cues used in the early order manipulation were (a) a valid Mode 1,3 IFF response, and (b) in the safe-passage corridor. The two hostile cues used in the early order manipulation were jamming and no IFF response.

The Patriot algorithm does not give aircraft points for not being in the safe-passage corridor or for never jamming or for stopping to jam friendly radar. These are the three cues in the middle column of the cue diagnosticity table. Interpretation of these cues is up to the operators. Previous research, both with the paper-and-pencil task and the Patriot simulator, had shown that most operators consider (a) not being in the safe passage corridor to be indicative of a hostile aircraft, and (b) "stopped jamming" to be indicative of a friendly aircraft. The former was the hostile cue, and the latter the friendly cue, used in the late order manipulation.

Finally, we checked the hard-copy printouts of the Patriot display prior to and during the experiment to ensure that all early and late order manipulations were implemented correctly. Correct manipulations required that only one new piece of information be present on the Patriot display between each data collection point. This manipulation was of particular concern because we found it extremely difficult to program the Patriot training simulator to implement all the tracks successfully in our previous study. The manipulation check indicated that all tracks were implemented correctly in the current experiment.

Hypotheses

We predicted that without the theoretically derived display, the early and late information order sequences would result in order effects consistent with the predictions of the Hogarth–Einhorn belief-updating model for both levels of initial information. That is, we predicted that participants would employ an anchoring and adjustment heuristic to process information so that they differentially weighed the most recent information based on their anchor for prior information. In contrast, we predicted that the theoretically derived display, which presented all information at all times, would foster a more global-processing strategy and, thereby, reduce the order effect.

RESULTS

Effect on Probability Estimates

Statistical analysis of the participants' probability estimates was performed for (a) data collection point C, which is after the early order manipulation, and (b) data collection point E, which is after the late order manipulation. Each analysis is considered in turn.

Point C

The probability estimates were coded in the direction specified by the initial information. For example, a Probability (Friend) = 0.80 estimate for an aircraft whose initial information indicated that it was friendly was coded as 0.80. Similarly, a Probability (Friend) = .20 estimate for an aircraft whose initial information indicated that it was hostile was recoded as .80. This recoding is required because, if one uses a Probability (Friend) scale, the Hogarth–Einhorn model predicts different looking-order effects (i.e., "fishtails"), depending on whether the initial information indicates a hostile or friendly aircraft. As can be seen in Figure 7.4, the Disconfirm/Confirm order is predicted to result in a higher Probability (Friend) estimate if the initial information indicates a friendly aircraft. In contrast, the Confirm/Disconfirm order is predicted to result in a higher Probability (Friend) estimate if the initial information indicates a hostile aircraft. By recoding the probabilities in the direction of the initial information, the Disconfirm/Confirm order is always predicted to have the higher value according to the Hogarth–Einhorn model. In addition, probabilities were multiplied by 100 when the data were coded to avoid potential errors due to forgetting decimal points. Consequently, they should be interpreted as percentages.

A within-subjects 2 (early order) × 2 (initial information) × 2 (display) Analysis of Variance (ANOVA) was performed on the difference score for

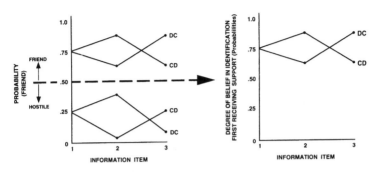

FIG. 7.4. Representation of the need to recode probability of friend estimates onto an "initial support" scale.

each participant's probability estimates between points C and A. Difference scores were used to control for differences in participants' starting probability estimates at point A. It should be remembered that because the probabilities were multiplied by 100, the differences can be greater than 1.0.

The ANOVA found two significant effects. First, there was a main effect for early order [$F(1,19) = 13.58$, $MSe = 119.71$, $p = .002$]. However, contrary to the predictions of the Hogarth–Einhorn model, the Confirm/Disconfirm order resulted in a higher mean C–A difference score than the Disconfirm/Confirm order: +4.09 versus −2.28, respectively, for a mean difference of +6.37 between the two orders. This finding indicates a primacy, not recency, effect.

Second, there was a significant early order × display interaction [$F(1,19) = 4.45$, $MSe = 64.34$, $p = .048$]. As shown in Figure 7.5, the theoretically

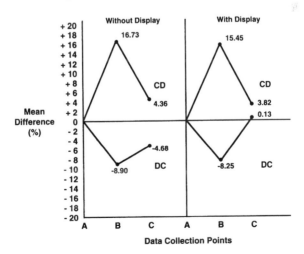

FIG. 7.5. Representation of early order by display interaction.

derived display significantly reduced the size of the mean difference between the two orders. The difference between the two orders was +9.04 without the display and +3.69 with it. This finding supports our predictions and suggests that early in the tracks' history, participants were using a more global-processing strategy with the display.

Point E

A within-subjects 2 (early order) × 2 (late order) × 2 (initial information) × 2 (display) ANOVA was performed on the difference score for each participant's probability estimates between points E and C. Early order was included in the ANOVA in order to test for effects involving early order × late order interactions at point E. No other early order effects were considered, for they were best represented at point C. Difference scores were used to control for differences in participants' starting probability estimates at point C.

The ANOVA found four significant effects. First, there was a main effect for late order [$F(1,19) = 19.13$, $MSe = 4906.28$, $p = .0003$]. The mean E – C difference for the Confirm/Disconfirm condition was significantly higher than that for the Disconfirm/Confirm condition: +5.86 versus −1.97, respectively, for a +7.84 difference between the two orders. This order effect again indicates a primacy effect and is contrary to the predictions of the Hogarth–Einhorn model.

The second effect was an early order × late order interaction [$F(1,19) = 6.00$, $MSe = 175.82$, $p = .024$]. As shown in Figure 7.6, the mean difference between the two orders was significantly larger when the early order sequence was Disconfirm/Confirm rather than Confirm/Disconfirm. The dif-

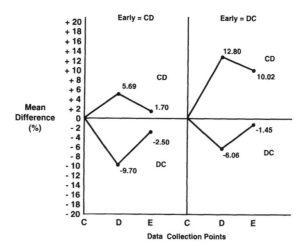

FIG. 7.6. Representation of early order by late order interaction.

ference between the two orders was +4.20 for the former and +11.47 for the latter. This was due to the substantially larger E − C difference for the Disconfirm/Confirm + Confirm/Disconfirm early and late order sequence.

The third significant effect was a main effect for initial information [$F(1,19)$ = 29.75, MSe = 1747.93, p = .0001]. There was a large, positive mean E − C difference (+14.69) for tracks whose initial information was hostile. This means that the participants were more convinced that these tracks were hostile aircraft by point E than C. In contrast, there was a large, negative mean E − C difference (−10.80) for tracks whose initial information was friendly. This means that the participants were less convinced that these tracks were friendly aircraft at point E.

The last effect was a late order × initial information interaction [$F(1,19)$ = 4.33, MSe = 171.01, p = .0511]. Figure 7.7 shows this effect and, more generally, helps explain all other significant E − C differences. The left panel shows the E − C mean differences for the tracks whose initial information indicates a hostile aircraft for both ordered sequences. The right panel shows the differences for the tracks whose initial information indicates a friend. The double hashed lines (// or \\) indicate whether the track left the safe-passage corridor between points C and D or points D and E. The only other piece of information was that the aircraft stopped jamming friendly radar.

Examination of the left panel in Figure 7.7 shows that, as expected, participants treated "Left Safe Passage Corridor" as a hostile cue; the differences were positive for both orders when the initial information indicated a hostile aircraft. And, as expected, participants treated "Stopped Jamming" as a friendly cue, but only when it preceded "Left Safe Passage Corridor"; the D − C mean difference was −4.12 for the Disconfirm/Confirm late order

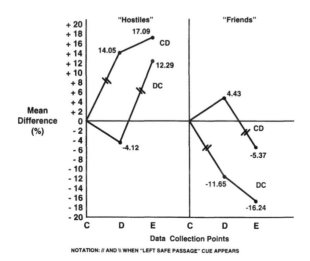

FIG. 7.7. Representation of late order by initial information interaction.

sequence. Contrary to expectations, "Stopped Jamming" was treated as a hostile cue when it followed "Left Safe Passage Corridor"; the E − D difference for the Confirm/Disconfirm order was +3.04.

Examination of the right panel in Figure 7.7 shows that the same thing happened when the initial information indicated that the track was friendly. "Left Safe Passage Corridor" was treated as a hostile cue; the differences were negative, as expected, when the initial information indicated a friend. "Stopped Jamming" was treated as a friendly cue when it preceded "Left Safe Passage Corridor"; the mean D − C difference was +4.43 for the Confirm/Disconfirm order. However, it was treated as a hostile cue when it followed "Left Safe Passage Corridor"; the mean E − D difference was −4.59 for the Disconfirm/Confirm order.

As shown in Figure 7.7, the late order × initial information interaction occurred because the difference between the two orders was significantly larger when the initial information indicated a friend than a hostile, that is, +10.90 versus +4.80, respectively. The main effect for initial information occurred, in part, because "Stopped Jamming" was treated as a hostile cue when it followed "Left Safe Passage Corridor." It is quite possible that the initial information main effect would have been observed even if "Stopped Jamming" always was treated as a friendly cue. But the positive E − C difference for tracks whose initial information indicated a hostile aircraft, and the negative E − C difference for tracks whose initial information indicated a friend, were accentuated because "Stopped Jamming" was considered a hostile cue when it followed "Left Safe Passage Corridor."

Examination of Figures 7.5 and 7.7 suggests that the early and late order primacy effects had different causes. The early order primacy effect was caused by the overweighing of the first piece of information in the ordered sequence (item 2 overall). The second piece of information in the sequence always moved the probabilities back in the predicted direction, just not as far as hypothesized by the Hogarth–Einhorn model.

In contrast, the late order primacy effect was caused by the reinterpretation of the meaning of the "Stopped Jamming" cue when it followed the "Left Safe Passage Corridor" cue. The reinterpretation increased the size of the positive E − C difference for the Confirm/Disconfirm order when the initial information indicated a hostile. Conversely, the reinterpretation increased the size of the negative E − C difference for the Disconfirm/Confirm order when the initial information indicated a friend. The joint effect of the reinterpretations was that the mean E − C difference for Confirm/Disconfirm late order was significantly different from the mean E − C difference for the Disconfirm/Confirm order.

Two sequences appear to have a part of the "fishtail" (i.e., recency) effect predicted by the model: (a) the Disconfirm/Confirm sequence when the initial information suggests a hostile, and (b) the Confirm/Disconfirm sequence when the initial information suggests a friend. An analysis was performed on the probabilities to assess whether these data were consistent with the predictions of the Hogarth–Einhorn model. Specifically, according to the model, confirming information should have a greater impact the lower the anchor. Consequently, when the initial information indicated a hostile aircraft, "Left Safe Passage Corridor" should have had a larger E – C difference for the Disconfirm/Confirm rather than Confirm/Disconfirm order. Varying over both levels of early order and display, this prediction was confirmed only two of four times.

Conversely, disconfirming information should have a greater impact the higher the probability anchor. Consequently, when the initial information indicated a friend, "Left Safe Passage Corridor" should have had a larger E – C difference for the Confirm/Disconfirm rather than the Disconfirm/Confirm order. This prediction was confirmed only once. These results further indicate a lack of support for the Hogarth–Einhorn model.

Effect on Identifications

The aforementioned analysis indicates that the order of information did affect participants' probability estimates. We now address whether it affected participants' identification judgments. A traditional chi-square test for independence could not be performed because each participant made an identification for each track; therefore, the entries in a chi-square table would not be independent. A more appropriate analysis is the chi-square test for marginal homogeneity (see, for example, Bishop, Fienberg, & Holland, 1975, Chapter 8). It tests the relationship between the distribution of responses (identifications) for each characterization of the independent variables.

It was not possible to estimate parameters for either the fully saturated four-way model or the three-way submodels. This occurred because the 20 participants did not provide sufficient variation in their responses for estimation of models with this many variables. Consequently, we focused on the two-way submodels. In particular, we focused on the four late order × initial information submodels in order to assess whether the results for the identification judgments tended to correspond to those for the E – C differences reported earlier. Each set of four two-way models was created by holding the levels on the other two variables constant.

Table 7.1 presents the results of the four late order × initial information models. We focus on two findings. First, the distributions of identifications for tracks whose initial information was hostile were significantly different than the distributions for tracks whose initial information was friendly for all four submodels. More importantly, the distributions were more skewed toward hostile identifications for the tracks whose initial information was hostile than they were skewed toward friendly identifications for tracks

TABLE 7.1
Late Order × Initial Information Results for Identifications

Display	Early	Initial Info (II)	LATE = CD			LATE = DC			x^2 and P-VALUES		
			IDs			IDs			Late (L)	Initial Info (II)	L × II
			H	U	F	H	U	F			
Without	CD	FRIEND	2	10	8	4	10	6	$x_1^2 = .520$	$x_1^2 = 24.22$	$x_1^2 = .750$
		HOSTILE	13	4	3	12	6	2	p = .770	p = .000	p = .686
Without	DC	FRIEND	2	9	9	5	11	4	$x_1^2 = 1.760$	$x_1^2 = 19.760$	$x_1^2 = 14.90$
		HOSTILE	15	3	2	9	9	2	p = .414	p = .000	p = .001
With	CD	FRIEND	4	7	9	4	7	9	$x_2^2 = 3.68$	$x_2^2 = 35.14$	$x_1^2 = .34$
		HOSTILE	15	3	2	13	5	2	p = .159	p = .000	p = .844
With	DC	FRIEND	4	9	7	8	7	5	$x_1^2 = 7.34$	$x_1^2 = 37.52$	$x_1^2 = 9.41$
		HOSTILE	17	2	1	9	8	3	p = .025	p = .000	p = .009

whose initial information was friendly. These results are consistent with those for the probabilities that the participants were more convinced that the former tracks were hostiles than the latter tracks were friends.

Second, there was a significant late order × initial information interaction when the early order sequence was Disconfirm/Confirm, both without and with the display. This occurred because there were more hostile identifications for (a) the Disconfirm/Confirm rather than Confirm/Disconfirm order for tracks whose initial information indicated a friend, and (b) the Confirm/Disconfirm rather than Disconfirm/Confirm late order for tracks whose initial information indicated a hostile. In both cases the "Stopped Jamming" followed the "Left Safe Passage Corridor" cue. The interaction was not significant for the Confirm/Disconfirm early order sequence.

These results are consistent with the significant early order × late order interaction shown in Figure 7.6. This is better illustrated by Figure 7.8, which shows the E – C probability differences for the late order × initial information interaction for each level of early order. As can be seen, the difference between the two late order sequences is significantly larger for the Disconfirm/Confirm early order sequence for both levels of initial information.

These results indicate a relationship between the participants' probability estimates and identification judgments. More importantly here, they indicate that information order affected the participants' identification judgments as well. However, the fact that there was only one late order main effect for the four two-way models suggests that the order effect was not as strong for the identification judgments.

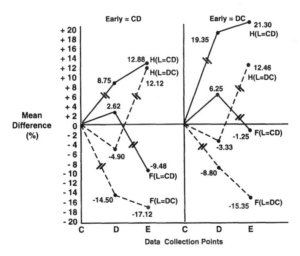

FIG. 7.8. Representation of late order by initial information interaction for each level of early order.

Effect on Engagements

Marginal homogeneity tests were performed on the engagement decisions. Again, we were unable to estimate model parameters for the four-way and three-way models because there was insufficient variance in participants' response patterns for the tracks to invert the covariance matrices. Therefore, we again focused on the four late order × initial information submodels.

The results are presented in Table 7.2. Again, there were four main effects for initial information. Hostile initial information resulted in significantly more engagements than friendly initial information, just as it resulted in more hostile identifications and lower Probability (Friend) estimates. In addition, there was a late order main effect for the Disconfirm/Confirm early order sequence, both with and without the display. For these conditions, there were more engagements for the Confirm/Disconfirm than Disconfirm/Confirm late order, consistent with the late order primacy effect for the probability estimates. In both cases, this was due to a significant late order × initial information interaction.

Consistent with the results for the probabilities and identifications, there were more engagements for (a) the Confirm/Disconfirm late order when the initial information indicated a hostile aircraft, and (b) the Disconfirm/Confirm late order when the initial information indicated a friend for three of the four submodels. The Confirm/Disconfirm early order (with display) condition was the only condition for which the interaction was not significant.

Although there were a number of significant effects, examination of Table 7.2 also shows substantial individual differences in participants' engagement decisions. The most extreme case was the "with display, Disconfirm/Confirm early order, initial information = hostile, and Confirm/Disconfirm late order" track. Eleven participants engaged this track, but nine did not. Yet all 20 participants saw exactly the same information. In addition, there was substantial disagreement among some participants for almost all 16 tracks. At one extreme, five of the participants did not engage a single track. At the other extreme, 5 participants engaged 6 or more tracks, 1 engaged 13 tracks, and another 14. Large individual differences also existed for the identifications (Table 7.1), particularly for certain tracks and for the probabilities. Large individual differences were observed in both previous studies as well.

DISCUSSION

The order in which the same information was presented to experienced Patriot air defense officers significantly affected their probability estimates, identification judgments, and engagement decisions. The order effects were obtained for eight tracks presented individually on the training simulator at

TABLE 7.2
Late Order × Initial Information Results for Engagements

| | | | # of ENGAGEMENTS | | x_1^2 and P-VALUES | | |
| | | | | | | Initial | |
Display	Early	Initial Info (II)	LATE = CD	LATE = DC	Late (L)	Info (II)	L × II
Without	CD	FRIEND	1	3	$x_1^2 = 1.05$ $p = .305$	$x_1^2 = 5.81$ $p = .016$	$x_1^2 = 5.81$ $p = .016$
		HOSTILE	7	3			
Without	DC	FRIEND	2	4	$x_1^2 = 13.33$ $p = .000$	$x_1^2 = 14.29$ $p = .000$	$x_1^2 = 16.36$ $p = .000$
		HOSTILE	13	3			
With	CD	FRIEND	1	3	$x_1^2 = 1.05$ $p = .305$	$x_1^2 = 7.27$ $p = .007$	$x_1^2 = 1.05$ $p = .305$
		HOSTILE	6	6			
With	DC	FRIEND	1	2	$x_1^2 = 9.41$ $p = .002$	$x_1^2 = 11.11$ $p = .001$	$x_1^2 = 20.00$ $p = .000$
		HOSTILE	11	2			

Fort Bliss, TX. Previous research showing order effects for air defense operators' judgments had used paper-and-pencil tasks (Adelman, Tolcott, & Bresnick, 1993) or a smaller number of tracks on the simulator because of implementation problems (Adelman & Bresnick, 1992).

Considering the probability estimates first, there was a significant main effect for the early order and late order variables. In both cases, the Confirm/Disconfirm sequence resulted in probabilities that were more in the direction of one's initial hypothesis than the Disconfirm/Confirm sequence. These results represent a primacy effect. They contradict the recency effect predictions of the Hogarth–Einhorn model that were obtained in previous research with air defense operators. They also are counter to previous research showing a recency effect in other domains, for example, auditing (Ashton & Ashton, 1988, 1990).

The early and late order primacy effects represented different information processes. The data suggest that the early order primacy effect represented an anchoring and adjustment heuristic where, in contrast with the Hogarth–Einhorn model, initial information was overweighed instead of underweighed compared to more recent information. Tolcott, Marvin, and Lehner (1989) found support for such a model in an Army situation assessment task. Adelman, Tolcott, and Bresnick (1993) posed it as an alternative to the Hogarth–Einhorn model, but failed to find support for it.

In contrast, the late order primacy effect was caused by the reinterpretation of the "Stopped Jamming" cue. "Stopped Jamming" was interpreted as indicative of a friendly aircraft, as predicted, before observing that the aircraft "Left Safe Passage Corridor." However, it was reinterpreted as indicative of a hostile aircraft when it followed "Left Safe Passage Corridor." Our explanation for this reinterpretation is that participants considered the latter sequence to be more representative of a hostile aircraft's pattern of behavior; in particular, where a hostile aircraft turns off its radar after leaving the corridor and approaching firing range in order to decrease the probability of detection by friendly radar. Although this explanation is post hoc, it does represent a viable script of enemy behavior.

More importantly from a theoretical perspective, information reinterpretations are not accommodated by the Hogarth–Einhorn model other than by "implicitly assum[ing] that the outcomes of the coding process already include whatever conditioning the subject has done based on prior evidence" (Hogarth & Einhorn, 1992, p. 39). Such an assumption preserves the model's "function-relation seeking" representation, that is, the perspective that "people's judgments are fundamentally based on relations between readily perceived cues (datum) and distal objects or events" (Hammond, 1988, p. 10). An alternative theoretical perspective, however, is that the participants were exhibiting "pattern seeking." That is, they were looking for the similarity or "representativeness of a data configuration to some a priori template that the data configuration should represent" (p. 39). The enemy behavior script described earlier represents such a data configuration. The key idea is that the resulting judgment and decision are not the function of an anchoring and adjustment heuristic that differentially weighs the importance of recent (or prior) information, but a pattern-matching process that attempts to explain the particular sequence of information.

There is no way to decide between the two alternative theoretical perspectives on the basis of the data presented herein. It is our post-hoc hypothesis that function–relation-seeking behavior was dominant early in the tracks' history, and pattern-seeking behavior late in the tracks' history. Such a hypothesis is consistent with Hammond's (1988, p. 7) more general observation that any task can induce "either pattern seeking or functional-relation seeking at any moment in time." It also helps explain the results with the display. Specifically, (a) the display was effective in inducing more global processing and, consequently, reducing order effects early in the tracks' history because participants were using a functional–relation-seeking approach; in contrast, (b) it was ineffective in dealing with order effects late in the tracks' history because they were employing pattern seeking. Again, we emphasize that this explanation is post hoc. We know of no research testing the effectiveness of different types of displays for global and pattern-seeking strategies. However, a number of studies have shown that performance with different displays is mediated by users' information-processing strategies (e.g., Adelman, Cohen, Bresnick, Chinnis, & Laskey, 1993; Jarvenpaa, 1989; Johnson et al., 1988).

There were three other significant effects for the latter half of the tracks' path. First, there was a significant initial information main effect. Participants were more convinced that tracks whose initial information indicated they were hostile were, in fact, hostile. In contrast, they were less certain that tracks whose initial information indicated they were friendly were friendly. These results are counter to those obtained by Adelman, Tolcott, and Bresnick (1993), in which participants were more certain of tracks whose initial information indicated a friendly not hostile aircraft. This difference was caused, at least in part, by the reinterpretation of the "Stopped Jamming" cue when it followed "Left Safe Passage Corridor." In addition, it may have been due to experience differences. Eighteen of the 20 participants in the current study had been in Desert Shield/Desert Storm. Previous research was prior to the war. Combat experience with scud missiles may have made the participants focus more on identifying hostile aircraft in the experiment. Although this explanation is post hoc, it is consistent with our explanation for why "Stopped Jamming" was reinterpreted in the current experiment, but not in previous ones.

Although we were unable to obtain more than two (of 20) officers without combat experience in the current study because of scheduling conflicts, previous air defense research had shown experience to be a significant factor. Adelman, Tolcott, and Bresnick found officers to be generally less susceptible to order effects, and Adelman and Bresnick found officers with European experience to be less susceptible to them too. Type of experience has been shown to affect expert judgment in other areas as well, such as auditor decision making (Johnson, Jamal, & Berryman, 1991), medical decision making (Patel & Groen, 1991), and naval decision making (Kirschenbaum, in press). We also observed large individual differences in the current study, as in both previous studies with Army air defense personnel.

Second, we obtained a late order × initial information interaction. This occurred because the difference between the two orders was significantly larger when the initial information indicated a friend rather than a hostile. And, third, there was a significant early order × late order interaction. The difference between the two late order sequences was significantly larger for the Disconfirm/Confirm early order sequence.

The effect on the probability judgments was reflected in the identification judgments and engagement decisions. Although there was not enough variation in participants' responses to estimate the four-way or three-way models, it was possible to test the four late order × initial information submodels. Consistent with the results for the probabilities, there were more hostile identifications and engagements for (a) the Confirm/Disconfirm late order when the initial information indicated a hostile aircraft, and (b) the Disconfirm/Confirm late order when the initial information indicated a friend. In both cases, the "Stopped Jamming" cue followed the "Left Safe Passage Corridor" cue.

Consistent with the early order × late order interaction for the probabilities discussed earlier, the identification and engagement differences between the (a) Confirm/Disconfirm late order + initial information = hostile and (b) Disconfirm/Confirm late order + initial information = friend conditions were largest, and principally significant, for the Disconfirm/Confirm early order sequence. There was only one (of four possible) late order main effects for the identifications, and only two for the engagements. This suggests that the effects for the probabilities translated into significant effects for the identifications and engagements only for the more extreme cases.

Taken together, the three studies with air defense personnel raise doubts about the generality of the Hogarth–Einhorn belief-updating model when information is presented sequentially. In the first experiment the recency effect was obtained for enlisted personnel, but not for officers. In the second experiment the recency effect was only obtained for three tracks whose initial information indicated they were friendly. Moreover, this was only for officers without European experience. Officers with European experience, who were more likely to see tracks with conflicting information rather than air defense personnel who had not left the United States, did not show the effect. In addition, an order effect was not obtained, either early or late in the tracks' history, for tracks whose initial information did not point toward friend or foe. Finally, in the current experiment, there were order effects, but no support for the recency effect predictions of the model. Moreover, there was minimal support for the model's assumptions about the effect of anchors of various strengths when the order effects appeared to have the predicted "fishtail" representation. In sum, the three studies raise doubts about the generality of the Hogarth–Einhorn model. However, they clearly illustrate that information order affects expert judgment.

It is our tentative conclusion that the order effects represent biased judgments. This conclusion may be premature, however, without data about the predictive accuracy of the judgments for different ordered sequences. For example, it is possible that the ordered sequence of "Left Safe Passage Corridor" and "Stopped Jamming" is more indicative of a hostile aircraft than the reverse order. In such a case, the order effect would be justified; it would not represent a bias. Such empirical data do not exist, however; nor are they likely to be available short of actual combat. The best one can do is examine training protocol and the expert judgment algorithm in the Patriot air defense system. According to both sources, order effects should not exist. Air defense officers are supposed to take into account all confirming and disconfirming information in the aggregate. Their judgments and decisions should not be affected by the sequential order of the information. However, all three experiments have shown that they are. This suggests an order effects bias, at least for the individual decision maker. Future research will study the two-person group representative of typical Patriot battery employment.

ACKNOWLEDGMENTS

The research described herein was supported by the U.S. Army Research Institute for the Behavioral and Social Sciences (ARI), Contract No. MDA903-89-C-0134 to Decision Science Consortium, Inc. (DSC). We would like to acknowledge the effort of the many people who contributed to this endeavor. In particular, we would like to thank Michael Drillings, Michael Kaplan, John Lockhart, and Michael Strub at ARI; and Andy Washko, Paul Richey, and participating Army air defense personnel at Fort Bliss, TX. The views, opinions, and findings contained in this chapter are those of the authors and should not be construed as an official Department of the Army position, policy, or decision, unless so designated by other official documentation.

Cognitive Redesign of Submarine Displays

Jill Gerhardt-Powals
Helene Iavecchia
Stephen J. Andriole
Ralph Miller III

This case study is based on many interactive, overlapping findings from cognitive science concerned with attention and memory; situational awareness; the way humans store, process, and retrieve knowledge; and other related issues. Not surprisingly, the overriding thesis of this case study is that if cognitive findings are used to develop a user interface, user performance will be improved.

The purpose of this chapter is to demonstrate that cognitively engineered user interfaces will enhance user–computer performance. This was accomplished by designing an interface using empirical findings from the cognitive literature. The interface was then tested and evaluated via a storyboard prototype.

HYPOTHESES

The following hypotheses were tested:

1. Human performance will be more accurate with the cognitively engineered interface.
2. Human performance will be faster with the cognitively engineered interface.
3. Workload associated with the cognitively engineered interface will be less.

194

4. Users will prefer the cognitively engineered interface.

The domain of this case study is Anti-Submarine Warfare (ASW). The interface designed for this study is for a decision-support system to be used by an Approach Officer during an Anti-Submarine Warfare Weapons Launch and Monitoring process consisting of two tasks: "Verify Firing Solution," and "Bring Ownship to Readiness."

THE COGNITIVE BACKDROP

The literature was examined to find the research most likely to impact the tasks. A number of specific areas were examined.

Situational Awareness

Situational awareness is a relatively new concern in human factors (Endsley, 1988; Fracker, 1988). Its origins are probably in aircrew jargon: "because they need it and sometimes lack it" (Taylor, 1989, p. 61). *Situational awareness,* as defined by HQ USAF AFISC/SE Safety Investigation Workbook, is "keeping track of the prioritized significant events and conditions in one's environment." Based on a study (Taylor, 1989) to investigate how aircrew understand "situational awareness," the following three broad domains of situational awareness were derived:

1. Demands on attentional resources (instability, complexity, variability)
2. Supply of attentional resources (arousal, concentration, division of attention, spare capacity)
3. Understanding of the situation (information quality, information quantity, familiarity).

According to Taylor (1989), by using the former domains we can see how situational awareness and decision making can be enhanced through thoughtful system design:

Control Demands on Attentional Resources. This can be accomplished by automation of unwanted workload, by fusing data, and by reducing uncertainty.

Improve the Supply of Attentional Resources. This can be achieved by prioritizing and cueing tasks to obtain the optimum attention-allocation strategy in accordance with mission goals and objectives and by organizing the structure of tasks to exploit the available resource modalities.

Improve Understanding. One method for improving understanding by design includes the presentation of information in cognitively compatible forms (3-D, voice, and pictorial multimodal displays).

Schematic Incorporation of New Information

New knowledge may be incorporated within existing schemas (Norman, 1982; Underwood, 1976). People make meaningful associations between new information and material stored generically in long-term memory (Bartlett, 1932; Chase & Ericsson, 1982; Posner, 1973; Schank & Abelson, 1977). Experienced operators extend metaphors from experience with other systems (Carroll & Thomas, 1982).

Based on these concepts, the display should be structured to aid integration of new information within existing memory structures and across displays (Woods, 1984). New information should be presented with meaningful aids to interpretation such as headings, labels, and verbal messages. The "system interface should be a coherent structure" (Carroll & Thomas, 1982, p. 22).

Context-Dependent, Reconstructive Retrieval

Accurate retrieval from memory is context dependent (Bobrow & Norman, 1975). Memories are reconstructed (Bartlett, 1932) in a recursive process: during retrieval, some contextual information about the to-be-recalled item is used to construct a partial description of the item (Norman & Bobrow, 1979). This description is developed, and the process continues recursively until the target item is recalled; this process is used, for example, when the names of former classmates are the target items (Williams & Hollan, 1981).

The implication of this cognitive characteristic (Murphy & Mitchell, 1986) is that names and labels used within displays should provide a meaningful context for recall and recognition. Display names should be conceptually related to function, and data should be grouped in consistently meaningful ways on the screen.

Top-Down and Bottom-Up Processing

Human information-processing system can be driven either conceptually (top-down) or by events (bottom-up); all input data make demands on processing resources, and all data must be accounted for; that is, some conceptual schema must be found for which the data are appropriate (Bobrow & Norman, 1975). In general, both conceptually driven and data-driven processing occur simultaneously with continuous top-down and bottom-up feedback until one of the candidate schemas is selected (Norman, 1986).

Because of this necessary human information processing, Murphy and Mitchell (1986) stated that designers should limit data-driven tasks to avoid consuming available processing resources. They made two additional suggestions:

1. Displays should include only that information needed by the operator at a given time and exclude peripheral information that is unrelated to current task performance.
2. Displays should provide the operator with the capability to reduce or increase the amount of displayed data.

Limited Processing Ability

The central high-level cognitive operations of attention, information processing, and memory have a limited processing capacity (Benbasat & Taylor, 1982; Bobrow & Norman, 1975; Lane, 1982; Shiffrin, 1976; Underwood, 1976). Only one high-level cognitive task can be performed at a time, with a slow switching rate for time sharing between tasks (Bobrow & Norman, 1975). The cognitive resources required for effective monitoring are subject to easy distraction and diversion to other mental tasks (Wickens & Kessel, 1981).

Because humans have a limited processing ability, Murphy and Mitchell (1986) made the following suggestions to use when designing user interfaces:

1. The user should not be required to monitor low-level, fragmentary information about the system or respond to isolated visual alert messages.
2. Alerts should be explicitly related to high-level, functional properties of the system.
3. Low-level information should be available for inspection in a form that presents it as inputs to higher level tasks.
4. Display design should aim to minimize the number of concurrent cognitive operations and allow the operator to concentrate on fault detection, diagnosis, or compensation.

Dynamic Internal Model

A system operator develops an internal model of the system's function and structure. This model is made up of dynamically interrelated knowledge structures (schemas) for general concepts, sequences of actions, situations, and system-related objects (Nisbett & Ross, 1980; Norman, 1982; Rumelhart & Ortony, 1977).

To promote this dynamic internal model, Norman (1982) suggested that display design should be internally consistent so that it reinforces the development and maintenance of an accurate system image. "All commands, feedback, and instruction should be designed to be consistent" (p. 105) with an accurate image of the system.

Inflexible Representation in Problem Solving

Information about the world is represented internally in multiple ways, including mental images that can be inspected and manipulated (Norman, 1980). In problem solving, people generally "tend to fixate on one form of representation even when multiple forms are available" (Rouse, 1983, p. 121).

The implicit result of this cognitive characteristic is that displays should provide the operator with more than one way of looking at data. In order to promote cognitive flexibility in problem solving, available displays should include high-level graphic overviews, but also allow the operator to view the same information in different formats and at different levels of detail (Murphy & Mitchell, 1986).

Cognitive Load

A good measure of task complexity or difficulty is the number of resources it uses (Moray, 1977; Kramer, Wickens, & Donchin, 1983; Berlyne, 1960; Sheridan, 1980; Welford, 1978; Hartman, 1961). This measure is known as *cognitive load*. The concept of cognitive load is important because it correlates directly with factors such as learning time, fatigue, stress, and proneness to error (Baecker & Buxton, 1987). Cognitive load must be kept in mind when designing user interfaces. Interfaces should be kept as simple as possible, using a minimal number of resources.

The characteristics of the inputs serving as load variables determine load effects. Meister (1976) listed the following major input characteristics that have been considered in the literature:

1. Number of channels through which inputs are presented.
2. Input rate (i.e., rate at which inputs enter the system and are presented to operators).
3. Stimulus-response relationships, particularly the complexity of these relationships.
4. Amount and type of information presented by the inputs.
5. The physical display that presents the input.

Although this research will be concerned with all of the listed input characteristics, the main thrust will be an emphasis on the physical display that presents the input. Because most of the inputs that load the system are visual, it is reasonable to suppose an interaction between load conditions and display characteristics (Meister, 1976).

Research has supported this theory. Poulton (1968) discovered that the probability of missing at least one target symbol in a simulated electronic display increased approximately linearly with number of targets and number of symbols present. Coffey (1961) found that performance was better with a lower density of alphanumeric material. Howell and Tate (1966) had similar results. Monty (1967) determined that performance in a keeping-track task was superior with numbers and letters to performance with symbols.

Interference

According to Baecker and Buxton (1987), degradation in the performance of one task due to competition for critical resources in another task is known as *interference*. Because many of the tasks that are performed using a computer have a high cognitive load, they are extremely susceptible to interference. Therefore, the designer should do whatever is within his or her power to reduce the likelihood of interference. This can be done by reducing the load associated with each task performance and by improving the quality of data available to the user (Baecker & Buxton, 1987).

An additional way to reduce interference is to take steps to minimize the possibility of competition. Different sensory modalities utilize different resources (Wickens, Sandry, & Vidulich, 1983). Therefore, when possible use of different sensory modalities will decrease interference.

Problem Solving

Two key points concerning problem solving, as described by Baecker and Buxton (1987), are first that it requires one's attention. It involves what is known as attentive behavior. Secondly, problem solving uses a relatively large number of resources. Consequently, problem solving is highly susceptible to interference. In working with a computer, there are two classes of problems that confront the user: operational and functional. Operational problems have to do with the *how* of performing work. Functional problems have to do with the *what* of that work (Baecker & Buxton, 1987).

Proper user interface design can help to minimize operational problem solving. Design features such as careful documentation and consistency are critical to this area.

Functional problems can also be helped by careful design. The user should be able to easily comprehend the situation using a minimum of valuable cognitive resources.

Skill Transfer and Consistency

The use of existing skills as the basis for performing new tasks is known as *skill transfer*. Four guidelines by Baecker and Buxton (1987) for successfully designing an interface to maximize this transfer are:

1. Build on the user's existing set of skills.
2. Keep the set of skills required by the system to a minimum.
3. Use the same skill wherever possible in similar circumstances.
4. Use feedback to effectively reinforce similar contexts and distinguish ones that are dissimilar.

By keeping the repertoire of skills small, the skills are used more often. And as the saying goes, "practice makes perfect." However, beyond the issue of practice is an underlying cognitive principle that if interfaces are consistent in the choice of skills to be required in a particular context, then the exploitation of in-system skills will be maximized (Baecker & Buxton, 1987).

Compatibility

According to Baecker and Buxton (1987), when the cause-and-effect behavior encountered in working with a system matches the users' expectations, it is said to have good stimulus–response (S–R) compatibility. Having good spatial congruence between items in a menu and the layout of function keys used for making menu selections is a good example of S–R compatibility.

In addition to spatial congruence, Baecker and Buxton described the other force that affects compatibility—custom. Much of compatibility is learned customs. As can be seen from the example of a toggle light switch, in North America the rule is generally "up is on, down is off." However, in many other countries, such as England, the convention is just the opposite. As a designer it is important to know the conventions and follow spatial congruency so that training time and operating load will be reduced (Baecker & Buxton, 1987).

Simplicity

Although simplicity is implicit in several of the cognitive attributes, it was felt that it deserved its own category. "A primary virtue of design is simplicity (Shneiderman, 1980, p. 255): from the medieval rule of Occam's Razor that states that the best explanation of physical principles is the simplest one, to the Bauhaus phrase "less is more," to the contemporary "small is beautiful."

Simple systems are easier to learn and easier to use. They are also easier to correct.

TESTING AND EVALUATION

Although this case study seeks to test and evaluate the *interface* of a computer-based information system, the literature review contains methods that have been used to evaluate the entire system, including the user interface. Historically, the interface has not been emphasized in the evaluation process as it has been in this research.

Evaluation of an information system is essential for making improvements and generating standard practices for future projects, but it is probably the most neglected activity of the information system cycle (Ahituv, Even-Tsur, & Sadan, 1986). One reason that it is neglected is because of the high cost of additional equipment and staff. However, the information system organization must weigh this cost against the potential price of resource ineffectiveness (Ameen, 1989).

In a study done by Hirschheim (1985), he interviewed 20 individuals in eight organizations who had recently implemented a new MIS (management information system). It was found that most organizations did not carry out formal postimplementation evaluation of their systems. Their most frequently cited reason was that such evaluations are time-consuming and require a substantial amount of resources. In one particular organization, formal evaluation is never done except when there is a high degree of dissatisfaction with the system or if the cost of the project significantly exceeds the allocated budget.

Although quite often MIS are not formally evaluated, occasionally a researcher studies MIS effectiveness as a dependent variable, or a practicing manager investigates the quality of the MIS being used. While acknowledging the importance of economic analyses of MIS value, researchers responded to the shifting emphasis from efficiency to user effectiveness by focusing either on MIS usage or user-perceived effectiveness (Srinivasan, 1985). If efficiency is emphasized over effectiveness, costs may decrease and utilization and throughput may increase, but quality and timeliness may suffer (Ameen, 1989).

The MIS usage approach uses behavioral indicators as surrogates for MIS effectiveness. Examples of such indicators are the number of reports generated, the number of changes made to a file, connect time, and so on. The perceived effectiveness approach uses measures of effectiveness as perceived by users of the system. Examples of such measures include user satisfaction, perceived system quality, and so on (Srinivasan, 1985).

Probably the most widely used indicator of MIS effectiveness has been user satisfaction (Hamel, 1988), and the most frequently used scale of user satisfaction was developed by Bailey and Pearson (1983) based on a review of 22 studies of user–computer interfaces. From this review, which was supplemented by interviews with DP professionals, the authors identified 39 factors relating to user satisfaction such as accuracy, reliability, timeliness, and confidence in the system. User satisfaction was defined as the weighted sum of users' positive or negative satisfaction with these 39 factors.

The instrument was tested and validated on a sample of 32 managers. The average reliability across the 39 factors was .93. Predictive validity was demonstrated by comparing questionnaire responses with a self-evaluation of perceived satisfaction obtained during the interview with the 32 managers; the resulting correlation was .79. For construct validity, the authors averaged the scores on the importance scale for each factor, rank ordered them, and correlated these rankings with self-assessed rankings which yielded a rank-order coefficient of .74. Taken as a whole, Bailey and Pearson concluded these efforts represented a successful translation of their definition of user satisfaction into a valid operational measure (Hamel, 1988).

Another measure of user satisfaction was by Jenkins and Ricketts, which was well grounded in a widely accepted theoretical model (Srinivasan, 1985). In order to uncover the underlying factors that constituted overall satisfaction with the system, Jenkins and Ricketts hypothesized that Simon's paradigm for the problem-solving process (intelligence, design, and choice phases) was an appropriate perspective of the manner in which users evaluate their experiences with the system. Using a factor-analytic approach to empirically test their claim, they postulated that there are five key underlying dimensions that make up overall user satisfaction: (a) Report Content, (b) Report Form, (c) Problem Solving, (d) Input Procedures, and (e) Systems Stability (Srinivasan, 1985).

Ives, Olson, and Baroudi (1983) offered a replication and expansion of the work of Bailey and Pearson (1983) by developing a shortened version of the scale. Factor analysis shrunk the number of factors from 39 to 32.

Raymond (1985, 1987) used a 20-item questionnaire, adapted from Bailey and Pearson's questionnaire. It excluded items deemed to be not applicable in a small business context.

The remaining instruments used to assess user satisfaction are mostly single items scales that have not been replicated elsewhere. For instance, Rushinek and Rushinek (1986) identified 17 independent variables that virtually showed no overlap with the sources from which Bailey and Pearson drew their items.

This case study adopts a multifaceted approach to testing and evaluation that was developed by Adelman (1991). He called for the integration of subjective and objective information and judgments using the application of several methods. Adelman suggested objective methods for testing performance criteria such as decision quality and speed and subjective methods for obtaining the users' opinions about the system's strengths and weaknesses.

In addition to the former methods, because it is hoped that a "cognitively friendly" interface will have a lower workload than one that is not, two validated subjective workload tools will be used. Hart (1987) defined workload as a multidimensional concept composed of behavioral, performance, physiological, and subjective components resulting from interaction between a specific individual and the demands imposed by a particular task.

NASA TLX (Task Load Index) is a multidimensional rating procedure for the subjective assessment of workload, which will be used in this study. It was designed to be used immediately following a performance of a task. Immediate procedures have been recommended as protection against information lost from short-term memory (Reid & Nygren, 1988). It is also an absolute measure meaning that it requires independent assessments of each experimental condition on abstract scale dimensions.

SWORD (Subjective WORkload Dominance), which had the Analytic Hierarchy Process (AHP) as a precursor (Saaty, 1980), is a different type of workload assessment that will also be used in this study. One difference in the SWORD approach, as compared to TLX, is that it uses completely retrospective evaluations: All comparisons are made after the rater has done all of the tasks. Vidulich and Tsang (1987) found that retrospective ratings had greater test–retest reliability than did immediate ratings. Also, SWORD has a redundant element; each task is compared with all other tasks.

RESEARCH MODEL DESIGN

A cognitively engineered user interface was designed and demonstrated via prototyping. An experiment was then conducted to measure user–computer performance while using the interface. According to Adelman (1992), one generally thinks of two kinds of experiments. The first tests the decision support system against objective benchmarks that often form performance constraints. The second kind of experiment, which was used in this research, is a factorial design in which one or more factors are systematically varied as the independent variables and the dependent variables are quantitative, objective measures of performance.

Adelman listed the following basic components of most factorial experiments: participants, independent variables, task, dependent variables (and measures), experimental procedures, and statistical analyses. The independent and dependent variables are discussed presently. The other four components are discussed in future sections.

Independent Variables

The design of this research was a 3 (interface) × 2 (problem scenario) × 2 (question complexity) factorial design. All three independent variables were within-subjects, meaning that each participant received every combination of interface, problem scenario, and question complexity.

Interface. The first independent variable—interface—included a baseline interface (Figures 8.1 and 8.2), a cognitively engineered interface as a result of this research (Figures 8.3 and 8.4), and an alternate-independent interface (Figure 8.5). All three interfaces were designed by different entities; however, all interfaces served the same purpose as a high-level command and control interface for use by an Approach Officer during an Anti-Submarine Warfare Weapons Launch and Monitoring task.

The baseline interface developed at Computer Sciences Corporation (Miller & Iavecchia, 1993) was designed to have features similar to those of typical military workstation displays. For example, there is a heavy reliance on alphanumerics, and information is often redundantly displayed. The alternate-independent interface was also developed by Computer Sciences Corporation.

Problem Scenario. The second independent variable—problem scenario—included the Bring Ownship to Readiness Problem and the Verify Firing Solution Problem. The Ownship Problem involved criteria that must be met concerning one's ownship which includes the proper speed, dive angle,

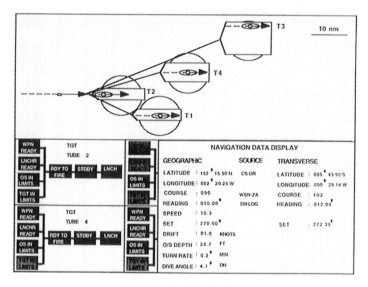

FIG. 8.1. Baseline display.

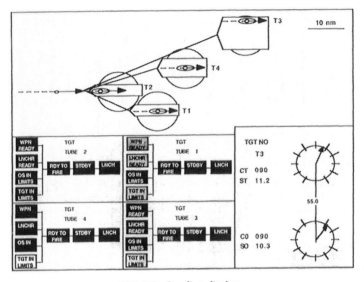

FIG. 8.2. Baseline display.

and turn rate. The Firing Problem involved criteria that must be met concerning the situational relationship between ownship and target, which includes ownship being aft the beam, in the baffles, and less than 40 nm away, as well as the torpedo envelopes properly covering the escape envelopes. Two different problem scenarios were used in this experiment to minimize the degree to which the results might be scenario dependent. Two different

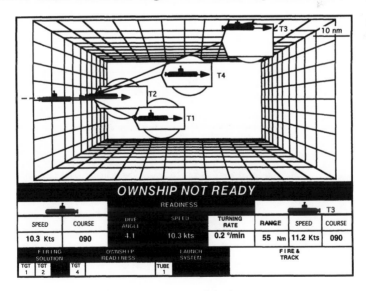

FIG. 8.3. Revised display.

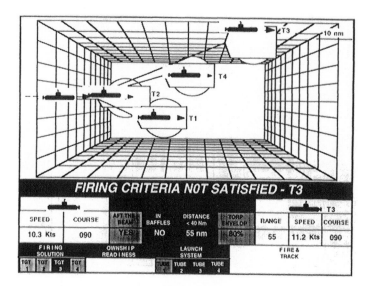

FIG. 8.4. Revised display.

problem scenarios were used in this experiment to minimize the degree to which the results might be scenario dependent.

Each participant completed eight trials in each of the two problem scenarios: Ownship and Firing. A trial had one or two components (i.e., questions), which leads to the third independent variable.

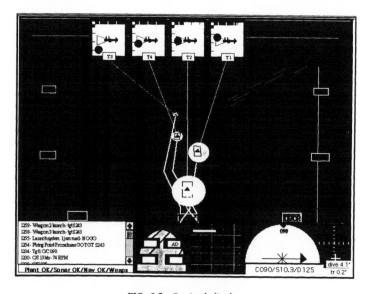

FIG. 8.5. Revised display.

Question Complexity. The third independent variable—question complexity—included a "yes or no" status question that has a *low* complexity and a "why?" question that results in a multiple-choice answer that has a *high* complexity. Initially for each trial within a problem, within an interface, a "yes or no" question was asked. If the response to the first question was "yes," then the trial was completed and the participant continued on to the next trial. If the response was "no" to the first question, then the participant was asked "why" the former response was "no." The participant was given a choice of answers from which to pick. After the participant answered the "why" question, the next trial began. By using two different levels of question complexity, internal validity was strengthened.

Dependent Variables

It was hypothesized that human performance would be more accurate and faster via the interface redesign. The workload associated with the interface was also expected to decrease. A final hypothesis was that the user would prefer the cognitively engineered interface.

In order to validate these hypotheses, both objective and subjective dependent variables were developed. Objective measures included accuracy and reaction time. Subjective measures included workload and preference.

Objective Measures. The objective measures—accuracy and reaction time—were captured in a computer program. As the participants interacted with each interface, their responses were recorded. Their response was compared to the correct response in order to calculate an accuracy measure. The amount of time that it took for them to respond was also recorded.

Subjective Measures. Workload was measured using two validated workload tools—SWORD and NASA-TLX—which are discussed in the evaluation section. SWORD was implemented using paper and pencil, whereas NASA TLX was automated on the computer.

Preference of interface was acquired through a question that asked participants to list the interfaces in order of their preference.

Conceptual and Operational Hypotheses

Specific hypotheses were derived by the translating of the cognitive findings into characteristics of the interface. Sometimes one cognitive finding suggested more than one conceptual hypothesis, which then indicated more than one operational hypothesis. In other instances, several findings implicated the same operational hypothesis.

Originally, there was to be a static experiment as well as a dynamic experiment. Because of time constraints, only the static experiment was possible. For that reason, all of the cognitive findings were not translated into interface characteristics. Some of them would have only been beneficial to an interactive display.

In the paragraphs that follow, the cognitive areas that were applicable to the static interface are noted. Conceptual and operational hypotheses derived from that cognitive area follow. The display design characteristic is indicated in the operational hypotheses.

Situational Awareness

Conceptual Hypothesis (CH) 1: If situational awareness can be improved by controlling demands on attentional resources, user–computer performance will be enhanced.

Operational Hypothesis (OH) 1.1: The automation of unwanted workload will result in better user–computer performance than when unwanted workload is not automated.

OH 1.2: The use of alert messages and color coding to reduce uncertainty will result in better user–computer performance than when uncertainty is not reduced by alert messages and color coding.

OH 1.3: Using appropriate data grouping to fuse data will result in better user–computer performance than when data are not appropriately grouped.

Schematic Incorporation of New Information

CH 2: If new information is presented together with meaningful aids to interpretation, user–computer performance will be enhanced.

OH 2.1: Presenting new information together with familiar metaphors such as miniature subs and the conventional use of "red" and "green" will result in better user–computer performance than when new information is not presented along with familiar metaphors.

Context-Dependent Reconstructive Retrieval

CH 3: If display names and labels provide a meaningful context for recall and recognition, user–computer performance will be enhanced.

OH 3.1: Display names and labels that are conceptually related to function will result in better user–computer performance than when display names and labels are not conceptually related to function.

OH 3.2: Data that are grouped in consistently meaningful ways on the screen will result in better user–computer performance than when data are not grouped in consistently meaningful ways.

Top-Down and Bottom-Up Processing

CH 4: By limiting data-driven tasks, user–computer performance will be enhanced.

OH 4.1: Using status indicators to limit data-driven tasks will result in better user–computer performance than when status indicators are not used to limit data-driven tasks.

CH 5: If informational load is minimized in order to allow human processing to comfortably absorb all displayed data, user–computer performance will be enhanced.

OH 5.1: Displays that include only that information needed by the operator at a given time will result in better user–computer performance than displays that include additional information.

OH 5.2: An interface with judicious redundancy will result in better user–computer performance than an interface with unnecessary redundancy.

Limited Processing Ability

CH 6: If low-level information is available for inspection in a form that presents it as inputs to higher level tasks, user–computer performance will be enhanced.

This conceptual hypothesis here will be operationalized in the same manner that was used for conceptual hypothesis 4.

Inflexible Representation in Problem Solving

CH 7: Displays that promote cognitive flexibility will enhance user–computer performance.

OH 7.1: Multiple coding of data will result in better user–computer performance than when data are only in one form.

Cognitive Load

CH 8: If cognitive load is decreased, learning time, fatigue, stress, and proneness to error will be decreased, thus user–computer performance will be enhanced.

This conceptual hypothesis will be operationalized in the same manner that was used for conceptual hypothesis 4.

Interference

CH 9: By reducing the cognitive load associated with each task performance and by improving the quality of data available to the user, the likelihood of interference occurring will be reduced and user–computer performance will be enhanced.

This conceptual hypothesis will be operationalized in the same manner that was used for conceptual hypothesis 4.

Problem Solving

CH 10: If the user can easily comprehend the situation using a minimal of valuable cognitive resources, functional problems will be mitigated, and user–computer performance will be enhanced.

This conceptual hypothesis will be operationalized in the same method that was used for conceptual hypothesis 1 and the same method that was used in conceptual hypothesis 4.

Simplicity

CH 11: If a system is simple, it is easier to learn and easier to use, therefore, user–computer performance will be enhanced.

With the exception of the operational hypotheses for conceptual hypothesis 7, all of the other operational hypotheses contribute to making the interface more simple and therefore would operationalize conceptual hypothesis 11 as well.

Definition of the Cognitive Design Principles

The operational hypotheses were used as a set of design principles in the development of the cognitively engineered interface. A succinct list of these 10 principles follows, with definitions and/or illustrative examples from the interfaces, using the same numbering system that was used in the operational hypotheses.

O.H.1.1 Automation of Unwanted Workload. This means eliminating calculations, estimations, comparisons, unnecessary thinking, and so on. For example, the cognitive interface demonstrates this feature with its use of red for dive angle and green for turning rate (see Figures 8.3 and 8.4). Red means the dive angle is out of the proper range, and green means that the turning rate is in the proper range. An example that does not use this feature is in the baseline interface (see Figures 8.1 and 8.2). In that interface one must look up the speed, which is 10.3, know the limit for speed, which is less than 9.5, make a mental comparison of the two, then come to the conclusion that 10.3 is out of range.

O.H.1.2 Alert Messages and Color Coding to Reduce Uncertainty. Uncertainty is reduced in the cognitive interface by both alert messages and color coding. For example, one can be certain that "ownship is not.ready." There is an alert message that says "OWNSHIP NOT READY," and it is also coded red. There is no way to be certain of ownship's readiness in the alternate interface (see Figure 8.5). One must know which criteria to be concerned with and must come to his or her own conclusion about ownship.

O.H.1.3 Appropriate Data Grouping in Order to Fuse Data. An example of this feature would be found in Figures 8.3 and 8.4. All of the firing criteria is grouped together in the center of the screen. It is further fused together by the alert message "FIRING CRITERIA NOT SATISFIED." In Figure 8.2, the firing criteria is found at the top left-hand side of the screen and the bottom right-hand side of the screen. There is no appropriate data grouping in order to fuse data.

O.H.2.1 Familiar Metaphors to Incorporate New Information. *Metaphor* is the use of an object or idea in place of another in order to suggest a likeness or analogy between them. In Figures 8.3 and 8.4, the familiar use of red and green is used to suggest "not satisfied" and "satisfied," respectively. Such use is consistent with conventional uses of the colors.

O.H.3.1 Display Names Should Be Conceptually Related to Function. The display name "FIRING SOLUTION" found in Figures 8.3 and 8.4 describes precisely the information contained in that section. The alternate interface (Figure 8.5) has a lack of display names conceptually related or otherwise.

O.H.3.2 Consistently Meaningful Grouping of Data. The data contained in the cognitive interface are both meaningfully grouped and consistently grouped. As seen in Figures 8.3 and 8.4, information concerning ownship is grouped under the blue sub, information concerning the target is grouped under the red sub, and information concerning ownship readiness is grouped under the heading "readiness." In Figure 8.4, one can see that the grouping is consistent.

This contrasts with both the baseline and alternate interfaces that do not have meaningful grouping of data. Ownship and firing criteria are found in more than one place and are interwoven with other types of data.

O.H.4.1 Use of Status Indicators to Limit Data-Driven Tasks. An example of this characteristic is found in the alternate interface (see Figure 8.5). At the bottom right-hand corner is a graph portraying dive angle and turning rate. If the cross is completely in the blue area, that indicates that the dive angle and turning rate are in the proper range. If the cross is above or below the blue, the dive angle is out of range; if the cross is to the left or right of the blue, that indicates that the turning rate is out of range. Therefore, the graph (i.e., the status indicator) limits the data-driven task of looking at the actual values of dive angle and turning rate and determining if they are within proper range.

O.H.5.1 Only Information That Is Necessary at a Given Time. Figure 8.3 is an example of only giving the information that is needed at a given time. Figure 8.3 has ownship criteria only, and Figure 8.4 has firing criteria only. This contrasts with Figure 8.5 in which both ownship and firing criteria are displayed at the same time.

O.H.5.2 Judicious Redundancy. Sometimes there is good reason for redundancy. One reason can be found in the next cognitive principle. However, it should not exist without a useful purpose.

Figure 8.2 is an example of an interface with unnecessary redundancy. There is no reason to show tube 3 and tube 4 in the illustrated situation. Also, there is no reason to have the box labeled "OS IN LIMITS" four times on the screen. It stands for "ownship within limits," and there is only one ownship.

O.H.7.1 Multiple Coding of Data to Promote Cognitive Flexibility. The cognitively engineered interface promotes cognitive flexibility by providing high-level graphic overviews and the same information in alphanumeric, color-coded, and alert forms (see Figure 8.3). The baseline and alternate interfaces generally displayed the data in only one form (Figures 8.1, 8.2, and 8.5).

Analysis of the Application of the Cognitive Design Principles in the Three Interfaces

An analysis of the application of these design principles to the three interfaces was conducted. For each condition (i.e., interface, problem, and question complexity combination), the use of the principle was reviewed.

If the design principle was completely absent, the condition received a 0. If the design principle was partially applied, the condition received a 1. If the design principle was fully applied, the condition received a 2. The sum of the conditions for baseline was 40; for alternate independent, 42; and for cognitively engineered, 78. A low sum indicates minimal use of the cognitive design principles.

All of this preparatory work led to the conduct of several experiments designed to measure—empirically and subjectively—the impact of the three interfaces and, especially, the impact of the cognitively engineered interface.

The next sections present the results of these experiments.

Experimental Design

As suggested earlier, the experiment involved three interfaces (i.e., the baseline interface, the cognitively engineered interface, and the alternate-independent), two problem scenarios (i.e., ownship and firing), and two question types (i.e., high complexity and low complexity). Because there were three interfaces, there were six different groups of participants in the experiment to counterbalance the presentation order of the interfaces as follows:

A,B,C A,C,B B,A,C B,C,A C,A,B C,B,A
(Each letter represents one of the three interfaces.)

By balancing the order in which the participants worked with the interfaces, an order effect was negated. By using a repeated measure (within subjects) design, in which each participant worked with each interface, one was able to partition out the error variance due to individual (vs. treatment) differences and, thus, increase the statistical power of the test (Adelman, 1992). Also, this design eliminated any selection bias typically found in between groups design, which is a threat to internal validity.

We met with the test participants before the first experimental day. The purpose of that meeting was to explain the domain (i.e., submarine warfare) and the workload tools (i.e., TLX and SWORD). Also, the assigning of groups took place at that preexperimental meeting.

The test participants consisted of 24 Drexel University students from the College of Information Studies. All had at least a minimum level of computer experience so that Macintosh training was not necessary.

The test was conducted in a computer lab at Drexel University equipped with 20 Macintosh computers with 13-inch color monitors. For the experiment a set of SuperCard screens was developed for each interface.

The experiment took place one day a week for three consecutive weeks. Each week the participant worked with a different interface. There was a week between each session to mitigate multiple-treatment interference, which is likely to occur whenever multiple treatments are applied to the same respondents (Campbell & Stanley, 1969).

Each participant spent approximately $1\frac{1}{2}$ hours at each session. This included interface training time and experimental data collection.

Control Variables

Environment, equipment, experimenter, and instrumentation were controlled for in this experiment. The environment and equipment were controlled because the experiment took place in the same room with the same machinery every week. The experimenter was controlled because the same experimenter taught every interface every week and oversaw the entire experiment.

The instrumentation was controlled in one of two ways. Either an instrument (e.g., a questionnaire) was consistently a part of the computer program, or it was on a printed sheet. Therefore, variables such as tone of voice, attitude, and so on, did not exist that could bias the data. No instrumentation involving observations or interviews was used, which could have caused a problem for internal validity (Campbell & Stanley, 1969).

Experiment Agenda

On the first day of the experiment, the domain and workload tools were reviewed. Then the experimental trial events were explained via view graphs. Basically, each trial involved: (a) reading a question, (b) pressing the return key when ready to begin, (c) viewing a screen from an interface, (d) pressing the return key when the answer is perceived, and (e) typing y for yes or n for no, or in the case of a multiple-choice answer, selecting an answer from a list and typing the number of the correct response.

After the trial events were thoroughly explained, the particular interface for the given session was taught through the use of view graphs and interaction between the researcher and participants. Questions were encouraged until all participants were comfortable with the interface.

The next step was to take the participants through a trial dry run by projecting the computer image onto a screen. Participation was encouraged, and once again questions were sought.

The final step of the training was for the participants to do a practice session on their own individual computers. There were two practice trials. After the practice, they had one last time to ask questions.

Once the training was complete, the experiment began. First each participant was given nine trials (the first trial was not used) concerning the Ownship Problem Scenario. Then they completed the TLX Workload Tool on Ownship.

Next they were given nine trials (the first trial was not used) concerning the Firing Problem Scenario. Then they completed the TLX Workload Tool on Firing. Finally, the participants completed some subjective questionnaires.

Evaluation

As mentioned in the review of the literature, this research adopted a multifaceted approach to testing and evaluation. Each interface was measured empirically and subjectively. As Adelman (1991) suggested, empirical methods were used for testing performance criteria, and subjective methods were used for obtaining the users' opinions about the interface's strengths and weaknesses. The following methods were used in this research.

Empirical Test

First the participant worked with the interface and answered empirical questions. Initially the answer was yes or no. However, if the answer was no, the participant had to then explain why. All of this was automated on the computer in the form of multiple-choice questions. In addition to each answer being counted either right or wrong, the amount of time taken to choose the answer was also calculated.

Subjective Questionnaire

A subjective questionnaire was completed by each participant for each interface. It was open-ended, allowing the participant to write whatever was on his or her mind.

SWORD Questionnaire. The SWORD (Subjective Workload Dominance) technique developed by Vidulich (1989) uses a series of relative judgments comparing the workload of different task conditions. After the participant completed all the tasks (i.e., in this experiment, three interfaces with two tasks each), he or she was presented with a rating sheet that listed all possible paired comparisons of the tasks (Vidulich et al., 1991).

Each pair of tasks appeared on one line on the evaluation form. One task appeared on the left side of the line, and another task appeared on the right. Between the two tasks were 17 slots that represented possible ratings. The rater marked the "equal" slot if both tasks caused identical amounts of workload. However, if either task has workload dominance over the other (i.e., caused more workload), the rater marked a slot closer to the dominant task. The greater the difference between the two tasks, the closer the mark was placed to the dominant task.

NASA-TLX (Task Load Index) Questionnaire. NASA TLX is a multidimensional rating procedure for the subjective assessment of workload. This technique was developed and evaluated at NASA Ames Research Center. The TLX provides an overall workload score based on a weighted average of ratings on six subscales: Mental Demand, Physical Demand, Temporal Demand, Performance, Effort, and Frustration (Selcon & Taylor, 1991).

TLX is a two-part evaluation procedure. First a subject is handed a paper with the 15 possible pairs of the six subscales. Subjects are instructed to circle the member of each pair that contributed more to the workload of that task. The second step is to obtain numerical ratings for each scale that reflect the magnitude of load imposed by each factor in a given task.

An overall workload score is computed by multiplying each rating by the weight given to that factor by each subject. The sum of the weighted ratings for each experimental task is then divided by 15 (i.e., the sum of the weights).

Research suggests (Byers, Bittner, & Hill, 1989) that the initial paired comparison sort procedure in TLX may be skipped without compromising the measure. A small study based on data collected for the pilot study for this research confirmed this. Therefore, in the final experiment for this research, the paired comparison sort procedure was skipped thus giving the participants less work to do. Further, the physical demand component was not included in the present study because static display screens were automatically presented during the experiment and the only motor response was to press a return key. For this experiment, TLX was automated on the computer, and the data were saved therein.

Preference Question. The interface preference question was presented at the conclusion of the experiment and was worded as follows: "Rank the three interfaces (by letter code) in their order of quality with quality defined as how well you could perform the task within each interface." The interfaces were coded by letters A, B, and C so that the test participants would not be biased by the use of the interfaces' experimental names (the coding scheme was A = cognitively engineered, B = baseline, C = alternate independent).

Research Sample—Discussion of External Validity

It was hoped that the research sample would consist of submariners. However, this was not possible. Instead it consisted of students from the graduate and undergraduate school of the College of Information Studies at Drexel University. They were males and females between the ages of 21 and 50 who have experience with computers.

Campbell and Stanley (1969) listed four factors that jeopardize external validity, which asks the question of generalizability. These factors are the following:

1. The reactive or interaction effect of testing.
2. The interaction effects of selection biases and the experimental variable.
3. Reactive effects of experimental arrangements.
4. Multiple-treatment interference.

Participants who took part in the pilot study did not participate in the actual experiment to avoid the reactive or interactive effect of testing that would jeopardize external validity (Campbell & Stanley, 1969). This effect can occur when a pretest increases or decreases the participant's sensitivity or responsiveness to the experimental variable and thus makes the results obtained for a pretest population unrepresentative of the effects of the experimental variable for the unpretested universe from which the experimental participants were selected.

The interaction effect of selection biases and the experimental variable (Campbell & Stanley, 1963) were not factors in this research due to the within-subjects design of the experiment and therefore not threats to external validity.

The reactive effects of experimental arrangements, which would preclude generalization about the effect of the experimental variable upon persons being exposed to it in nonexperimental settings, which is Campbell and Stanley's third threat to external validity, was somewhat difficult to control. The ideal would have been not to allow the participants to know that they were taking part in an experiment. This was not possible. In spite of that, the following was done to mitigate the reactive effects:

1. No one was forced to participate.
2. Participants were encouraged to do their best.
3. Mature young adults were the participants.
4. A comfortable atmosphere was provided.

The multiple-treatment interference, likely to occur whenever multiple treatments are applied to the same respondents because the effects of prior treatments are difficult to erase, is the final threat to external validity. To mitigate this factor, the three treatments were separated by one week. Also, the three treatments were given in six different arrangements to eliminate an order effect.

Although the domain of this study is submarines, that is not the emphasis of this research. This research is measuring the significance of having cognitively engineered interfaces. Because people (submariners or otherwise) tend to be similar cognitively (Dodd & White, 1980), it should not diminish the results of the study because nonsubmariners were used. It would have only been necessary to teach the subjects enough about the domain so that they were comfortable. Once that was accomplished, any results concerning the cognitively engineered interface would be generalizable to all people.

Instrumentation

Most of the instruments were described in the evaluation section. In addition to those already mentioned, there was a background questionnaire that was filled out by each participant and a posttest questionnaire that was completed after the pilot study was held and was also completed after the final experiment.

Pilot Study

A pilot study was conducted mainly to test out the procedure and the instruments. Although all three interfaces were not completed by the time of the study, a great deal of knowledge was acquired. Fifteen participants took part in the study. The most valuable information learned was that the experiment would take much longer than anticipated.

The posttest questionnaire gave insight into ways to make the final experiment better. This included more careful teaching of interfaces, clearer instructions on questionnaires, and allowing more time for various activities. Overall the participants seemed to have a positive experience.

Data Collection

As mentioned previously, an experiment took place comparing the baseline interface prototype with the cognitively engineered interface prototype and the alternate-independent interface prototype. Performance data and the TLX data were captured in the computer. In addition, as mentioned in the evaluation section, data were also collected via questionnaires. This included background information on the participants as well as subjective questionnaires concerning the interfaces.

ANALYSIS AND RESULTS

Reaction Time

The cognitively engineered interface averaged 1.33 sec per question as compared to the baseline interface, which averaged 5.58 sec, and the third interface, which averaged 4.06 sec. That is a 76% improvement over the baseline interface and a 67% improvement over the third interface. The cognitively engineered interface was faster, regardless of the problem scenario (i.e., ownship or firing) or the question complexity (i.e., low or high).

To analyze the performance data, raw cell means were calculated for each subject and condition. Then they were subjected to an Analysis of Variance (ANOVA) using the BMDP2V PROGRAM (Dixon, 1977). Post-hoc tests were conducted using simple effects analysis and the Tukey HSD (honestly significant difference). A significance level of .05 was applied to all analyses.

A 3 × 2 × 2 omnibus-repeated-measures ANOVA was conducted with Interface (baseline, cognitively engineered, and alternate independent), Problem (ownship, firing), and Question Complexity (low, high) as variables. The analysis found main effects for Interface [$F(2,46) = 126.11, p < .001$], Problem [$F(1,23) = 15.42, p < .001$] and Question Complexity [$F(1,23) = 121.18, p < .001$], as well as for the interaction of the Interface and Problem [$F(2,46) = 14.19, p < .001$] and the interaction of Interface and Question Complexity [$F(2,46) = 90.92, p < .001$].

An interaction for Problem and Question was not found [$F(1,23) < 1.0$], nor was there a three-way interaction of Interface, Problem, and Question Complexity [$F(2,46) < 1.0$]. The results for the reaction-time main and interaction effects are graphically illustrated in Figures 8.6 through 8.10.

Because there were more than two interfaces, the main effect for the Interface factor was probed using the Tukey HSD test. Significant differences were found for every interface combination. Reaction time to the cognitively engineered interface was faster than the alternate-independent design, which in turn was faster than the baseline.

The two-way interaction between Interface and Problem was probed using simple effects analyses and the Tukey HSD test. For the simple effects analysis, the error terms for the Interface main effect and the Interface by

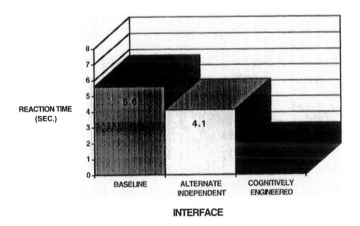

FIG. 8.6. Interface reaction times means.

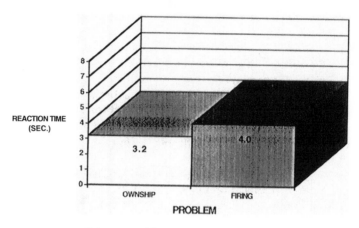

FIG. 8.7. Problem reaction times means.

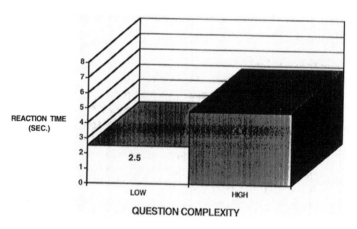

FIG. 8.8. Question complexity reaction times means.

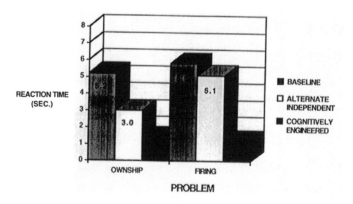

FIG. 8.9. Integration of problem and interface reaction times means.

220

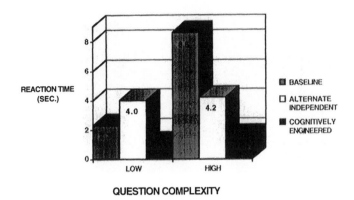

FIG. 8.10. Integration of question complexity and interface reaction times means.

Problem interaction were pooled to create a corrected mean-square error term. The degrees of freedom for the error term were also corrected using the Satterthwaite correction equation.

A one-way ANOVA (Interface: baseline, cognitively engineered, and alternate independent) was executed for each level of the Problem factor. For Problem = Ownship, an Interface main effect was found [$F(2,89) = 30.94$, $p < .001$], with the Tukey HSD revealing a significant difference between every interface pair. That is, reaction time to the cognitively engineered interface was faster than the alternate independent, which in turn was faster than the baseline.

For Problem = Firing, an Interface main effect was found [$F(2,89) = 50.89$, $p < .001$], with the Tukey HSD revealing a significant difference between cognitive and baseline as well as between cognitive and alternate effects. That is, for Problem = Firing, reaction time to the cognitively engineered interface was faster than both of the other interfaces (see Figure 8.11).

The two-way interaction between Interface and Question Complexity was probed using simple effects analyses and the Tukey HSD test. For the simple effects analysis, the error terms from the omnibus ANOVA for the Interface main effect and the Interface by Question Complexity interaction were pooled to create a corrected mean-square error term. The degrees of freedom for the error term were also corrected using the Satterthwaite correction equation.

A one-way ANOVA with Interface (i.e., baseline, cognitively engineered, and alternate independent) was executed for each level of the Question Complexity factor. For Question Complexity = Low, an Interface main effect was found [$F(2,92) = 15.37$, $p < .001$], with the Tukey HSD revealing a significant difference between every interface pair. That is, for low complex-

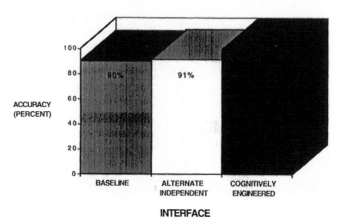

FIG. 8.11. Interface accuracy means.

ity, reaction time to the cognitively engineered interface was faster than the baseline, which in turn was faster than the alternate independent.

For Question Complexity = High, an Interface main effect was also found [$F(2,92) = 93.73$, $p < .001$], with the Tukey HSD again revealing a significant difference between every interface pair. That is, for high complexity, reaction time to the cognitively engineered interface was faster than the alternate independent, which in turn was faster than the baseline (see Figure 8.12).

Accuracy

The cognitively engineered interface averaged 99% accuracy as compared to the baseline interface, which averaged 90%, and the third interface, which averaged 91%. The cognitively engineered interface was 10% more accurate

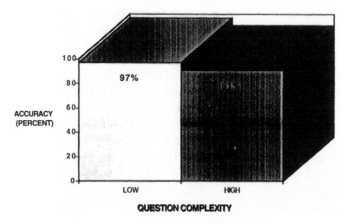

FIG. 8.12. Question complexity accuracy means.

than the baseline interface and 9% more accurate than the third interface. The cognitively engineered interface was more accurate regardless of the problem scenario (i.e., ownship or firing) or the question type (i.e., yes/no or multiple choice).

Using the accuracy data presented in Figure 8.12, a 3 × 2 × 2 omnibus-repeated-measures ANOVA was conducted with Interface (baseline, cognitively engineered, and alternate independent), Problem (ownship, firing), and Question Complexity (low, high) as variables. The analysis found main effects for Interface [$F(2,46) = 14.66$, $p < .001$], Question Complexity [$F(1,23) = 23.26$, $p < .001$], and for the interaction of Interface and Question Complexity [$F(2,46) = 6.02$, $p < .005$].

A main effect for Problem was not found [$F(1,23) < 1.0$], nor for the interaction of Interface and Problem [$F(2,46) < 1.0$], nor for the interaction of Problem and Question [$F(1,23) < 1.0$]. Finally, there was no three-way interaction of Interface, Problem, and Question Complexity [$F(2,46) < 1.0$]. The results for the accuracy main and interaction effects are graphically illustrated in Figure 8.13.

The main effect for the Interface factor was probed using the Tukey HSD test. A significant difference was only found between the cognitive interface and the baseline interface. The former was more accurate than the latter (see Figure 8.13).

The two-way interaction between Interface and Question Complexity was probed using simple effects analyses and the Tukey HSD test. Again, the error terms for the Interface main effect and the Interface by Question Complexity interaction were pooled to create a corrected mean-square error term. The degrees of freedom for the error term were also corrected using the Satterthwaite correction equation.

A one-way ANOVA (Interface: baseline, cognitively engineered, and alternate independent) was executed for each level of the Question Complexity

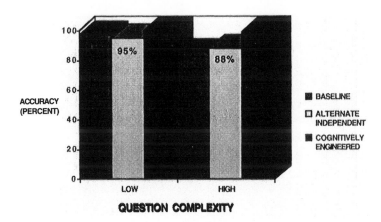

FIG. 8.13. Interaction of interface and question complexity accuracy means.

factor. For Question Complexity = Low, an Interface main effect was not found [$F(2,91)$ = 1.12, p > .25].

For Question Complexity = High, an Interface main effect was found [$F(2,91)$ = 9.59, p < .001], with the Tukey HSD revealing a significant difference between cognitive and baseline as well as between cognitive and alternative. That is, for high complexity, accuracy in the cognitively engineered interface was better than the alternate independent, and the cognitive was also better than the baseline.

Postexperiment Questionnaire

At the end of the 3-day experiment, a questionnaire was completed concerning the training of the participants in the three interfaces and in TLX and SWORD. One-hundred percent of the participants noted that the training of the three interfaces was sufficient. One-hundred percent also said the SWORD training was sufficient, and all but one said that the TLX training was sufficient.

When asked to rate the ease of use of TLX, 19 out of 24 gave it a 1 (very easy) out of a scale from 1 to 5. Three gave it a 2, one gave it a 3, and one gave it a 4.

When the participants were asked to rate ease of use of SWORD, only 7 out of 24 gave it a 1 (very easy), 8 gave it a 2, 8 gave it a 3, and 1 gave it a 4.

When asked if the tools were a good measure of workload, 19 responded in the affirmative for TLX, whereas only 13 responded affirmatively for SWORD. The main reason given for a negative reaction to SWORD was that it was too difficult to remember the interfaces from the previous weeks.

Preference

When the participants were asked in which order they preferred the interfaces, 20 out of 24 participants selected the cognitively engineered interface first. That represents 83% of the participants. The cognitively engineered interface was also tied for first twice and was rated as second twice.

When a Friedman Test was computed for these data at a .01 significance level, with degrees of freedom of 2 and 46, the null hypothesis was rejected. Therefore, we may conclude that there is a tendency to prefer one or more interfaces over the other(s).

For multiple comparisons, with $(23)*(2)$ = 46 df, t (.99) is found to be 2.704. The critical difference between the means was determined to be 12.168. Any two interfaces whose rank sums have a greater difference than the former may be regarded as unequal. Therefore, the cognitively engineered interface may be preferred to the baseline interface (difference = 33) and the alternate interface (difference = 28.5).

TLX

The test participants rated the cognitively engineered interface as having a relatively low workload compared to the other two interfaces. The workload totals were 45 for the cognitive interface, 137 for the baseline interface, and 124 for the alternate interface. Specifically, the workload experienced with the cognitively engineered interface was 67% less than the baseline interface and 64% less than the alternate interface.

To analyze the TLX data, raw cell means were calculated for each participant and each problem scenario (i.e., ownship and firing). Then they were subjected to an Analysis of Variance (ANOVA) using the BMDP2V PROGRAM (Dixon, 1977). Post-hoc tests were conducted using simple effects analysis. A significance level of 0.05 was applied to all analyses. Using the TLX data, an omnibus-repeated-measures ANOVA was conducted with Interface (baseline, cognitively engineered, and alternate independent) and Problem (ownship, firing) as variables. The analysis found main effects for Interface [$F(2,46) = 37.45, p < .0001$] and Problem [$F(1,23) = 23.37, p < .001$]. There was also an interaction effect of the Interface and Problem [$F(2,46) = 4.04, p < .03$].

Because there were three interfaces, the main effect for the Interface factor was probed using the Tukey HSD test. For a significance level of .01, a significant difference was found between the cognitive interface and each of the other two interfaces. Therefore, we may conclude that the cognitive interface had significantly less workload than the baseline interface and the alternate interface.

A one-way ANOVA (Problem: ownship and firing) was executed for each level of the Interface factor. For Interface = Baseline, a Problem main effect was found [$F(1,23) = 14.07, p < .01$]. That is, for the baseline interface, workload in the ownship problem was significantly less than in the firing problem.

For Interface = Alternate, a Problem main effect was also found [$F(1,23) = 5.72, p < .03$]. That is, for the alternate interface, workload in the ownship problem was significantly less than in the firing problem.

For Interface = Cognitively Engineered, a Problem main effect was not found [$F(1,23) = .05$]. Because F was less than 1, there is no significance. That is, for the cognitive interface, workload in the ownship problem was *not* significantly different than in the firing problem.

SWORD

As mentioned previously, when the participants were asked to rate the ease of use of SWORD, only 7 out of 24 gave it a 1 (very easy), 8 gave it a 2, 8 gave it a 3, and 1 gave it a 4. Also, when they were asked if the tools were a good measure of workload, only 13 responded affirmatively for SWORD. The main reason given for a negative reaction to SWORD was that it was too difficult to remember the interfaces from the previous weeks. For these reasons, SWORD is not included in the results.

SUBJECTIVE QUESTIONNAIRE

Participants completed a subjective questionnaire on each interface. It included two questions: What did you like and/or find most helpful about this interface? and What did you dislike and/or find most difficult about this interface?

Cognitively Engineered Interface

The response to the cognitively engineered interface included something positive from all 24 participants. Some participants had multiple likes or dislikes. Sixteen noted that the cognitively engineered interface enabled fast decisions. Four people mentioned that the color coding made it easy to use. Three commented that they liked the layout. One person mentioned that it was easy to learn; another liked the grouping of criteria. Other favorable comments included: "layout of firing criteria," "torpedo envelope was calculated," "eliminated need to memorize criteria," and "substatus always listed in the same order in the same place."

There were four participants that mentioned dislikes of the cognitively engineered interface. One participant did not like the arrangement, and one said that it had zero mental demand. Negative comments also included, "useful data at bottom of screen" and "so much information to tune out."

Baseline Interface

Twenty-two participants had something that they liked about the baseline interface, and 23 had something that they disliked. Twenty-one participants mentioned the use of red and green as a positive. Three liked the graphic nature, and two liked the layout. Other positives mentioned were "grid method," "ownship criteria easy to pick out," "relatively easy to evaluate criteria," and one spot for information.

Many of the 23 that had a dislike of the baseline interface had more than one dislike. Four participants noted that the information was too scattered. Five did not like the torpedo envelopes. Three did not like the changing of target numbers at the bottom right-hand side. Two participants each mentioned the following dislikes: "hard to judge which target to concentrate on," "judging of baffles," "display is confusing," and "random numbering of target tubes." There were 17 additional negative comments for the baseline display.

Alternate-Independent Interface

Twenty-three participants had something good to say about the alternate-independent interface, and 24 participants had something bad to say about it. Eleven mentioned something positive about the colors that were used. Ten liked the situational plot. Some of the other positive comments mentioned were: "interesting," "miles graphed out," "pleasing to the eye," "challenging," "aircraft-like control panel," "cross hair," "let you get involved," "made you think and analyze," and "micro display."

Out of the 24 participants that had something derogatory to say, 8 commented that calculating the envelope percentage was too difficult. Seven said that they disliked searching more than one place for the answer. Four commented that the microdisplay was too busy. Three participants mentioned that the target boxes were too small. Two people each noted the following: "having to look at the different pictures," "distance layout at extreme distance," and "lack of consistency with color." Other singular negative comments included: "speed hard to read," "labeling was not useful," "cluttered display," "overabundance of information," "redundancy," and "numbers too small."

INTERPRETATION OF RESULTS

Reaction Time

As detailed in the Data Analysis section, the cognitive interface was faster than the alternate interface, which was faster than the baseline interface.

To determine if there was also a correspondence between the Interface-by-Problem interaction associated with the reaction-time data and the design-principle usage, for the ownship case, the baseline had only a combined total of 19, which corresponds to its relatively poor reaction-time performance relative to the other interfaces. The alternate had a design principle score of 26 for ownship, and the cognitive interface had a score of 39 for ownship, which corresponds to their ownship reaction times.

To determine if there was also a correspondence between the Interface-by-Question-Complexity interaction associated with the reaction-time data and use of the cognitive-design principles, the totals for low- and high-complexity questions were lowest for the conditions with worst performance. Specifically, for the low-complexity question, the alternate independent had relatively poor reaction time performance compared to the other interfaces (Figure 8.13).

The high-complexity question, the baseline had relatively poor reaction-time performance.

The cognitively engineered interface had the best performance for both low- and high-complexity questions.

Accuracy

As detailed in the Data Analysis section, the cognitive interface was more accurate than the alternate interface, which was more accurate than the baseline interface. This was to be expected based on the utilization of the cognitive design principles. The cognitive interface had the highest application of the principles, the alternate independent was second, and the baseline was third.

Because accuracy was rather high in all three interfaces, comparing results with the design principles is not as compelling as it was for reaction time. Nevertheless, the cognitively engineered interface had the highest condition totals as well as the highest accuracy, regardless of problem scenario or question complexity.

Preference

The results of the subjective questionnaire help to explain why the participants preferred the cognitively engineered interface. Generally, it was easy to learn and easy to use. It also enabled fast decisions.

Another explanation for the preference of the cognitive interface would be that according to TLX it had a relatively low workload compared to the other two interfaces.

CONCLUSIONS AND IMPLICATIONS

This research attempts to broaden the methods, tools, and techniques that we use to design, develop, implement, and test user–computer interfaces. More specifically, the case study reports on some experiments designed to test the applicability (and potency) of empirical findings from cognitive science, findings that can provide insight into features that interfaces should possess to enhance the overall user–computer interaction process.

The hypotheses derived from cognitive science, as well as the data collected to confirm them, suggest that there may well be empirical relationships between cognitive science and user–computer interface design and user–computer interaction performance.

As with human factors research and research in ergonomics, we now have additional insight into possible sources of enhanced performance. The purpose of this research was to demonstrate the relationship between empirical findings in cognitive science and UCI design and performance. This has been achieved. It is now possible to extend the largely conceptual assertion that cognitive science can help UCI design and UCI performance. Countless articles and books have hailed the relationship as one with enormous potential (pointing at times to "disembodied" assertions like the use of icons and hierarchical data structures as "cognitive systems engineering"), but virtually none of them point to any empirical evidence. This research represents one of the first attempts to establish the viability of the relationship and lights the way for others to search for additional examples of how empirical findings from cognitive science can enhance user–computer interaction performance.

Specifically, based on this research, the cognitively engineered interface exhibited superior performance in terms of reaction time and accuracy measures, was rated as having the lowest workload, and was the preferred interface. Further, the cognitive-design-principles analysis indicated that the baseline and alternate interfaces minimally employed the design principles that were explicitly applied to the cognitively engineered interface. This suggests that the lack of applying these design principles could result in poor performance, whereas the application of these cognitive-design principles could enhance user–computer performance.

In addition, the results of the reaction time, accuracy, and workload data for the three interfaces were analogous in every case to the application of the cognitive design principles. That is, even when the alternate interface was compared to baseline for a particular situation, the interface with the higher application of the cognitive design principles performed better.

Contributions

One of the contributions of this research is a methodology that can be employed in designing a cognitively engineered interface. From the first step, which is reviewing the literature, to the final step of evaluation, the process is detailed.

Another contribution that is a result of this research is a set of 10 cognitive-design principles that could be used if a designer's goals include accuracy, speed, low workload, and a preference for the interface. The following 10 design principles should be applied:

1. Automation of Unwanted Workload
2. Use of Alert Messages and Color Coding
3. Appropriate Data Grouping

4. Use of Familiar Metaphors
5. Display Names Related to Function
6. Consistent Meaningful Grouping
7. Use of Status Indicators
8. Necessary Information Only
9. Judicious Redundancy
10. Multiple Coding of Data to Promote Cognitive Flexibility

A final contribution of this research is a methodology that can be used to conduct an experiment involving three interfaces. All the details have been worked out to assure internal and external validity.

Recommendations

It is recommended that in professional practice it is important to recognize that "To the User, the Interface Is the System" (Boyell, personal communication, October 1990). Further, it is important to realize that a cognitively engineered interface enhances human performance. Therefore, in designing an interface, an attempt should be made to make it cognitively friendly.

Further research is necessary to determine the impact of specific cognitive-design principles on performance. For example, one could develop and experimentally test several versions of the baseline design with each version incorporating a different cognitive-design principle. Such experimentation may lead to the identification of a set of critical cognitively engineered guidelines for display designers.

Also, it would be interesting to gain insight into the acceptance of the various interface designs (i.e., baseline, cognitive, alternate) in the targeted user community, which was submariners. Because of the portability of the experimental software (i.e., SuperCard, Mac II), a similar study to the one reported here could conceivably be conducted at a Navy submarine base. It may be determined that there are other important cognitive considerations, such as boredom, for example.

A final suggestion for further research is to apply the 10 cognitive-design principles to a completely different domain and determine if human–computer performance is enhanced.

ACKNOWLEDGMENTS

This case study was supported by the Integrated Systems Division of the Computer Sciences Corporation of Moorestown, NJ. Thanks to A. J. Alamazor, Mike Sullivan, and Brian Magee for their help with the project, especially with the programming of the tools that made the evaluation process so easy.

Thanks also to the Ben Franklin Technology Center of Southeastern Pennsylvania for its support of the Center for Multidisciplinary Information Systems Engineering (MISE Center), and to the students of Drexel University's College of Information Studies for their participation of willing and able students. Finally, we would like to thank Betty Jo Hibberd for her able support during the entire course of the project.

CHAPTER NINE

Issues, Trends, and Opportunities

This book has attempted to demonstrate the linkage between findings from cognitive science and user–computer interface design and development. Case studies were used as the demonstration/communications vehicles. We have tried to demonstrate that there is a measurable positive effect of adding interface features from what we know about human cognitive processes. We undertook the project because whereas there was a lot of discussion in the literature about "cognitive engineering," there was relatively little work devoted to demonstrating the process empirically.

But there are as many open questions now as when we began. In fact, we could easily argue that the field of UCI design and development is changing almost daily. There are questions surrounding the development and implementation of "intelligent" and "adaptive" interfaces, task allocation, questions surrounding de facto (and other) standards, and questions surrounding information technology solutions to improved user–computer interaction.

ADAPTIVE AND AUTOMATED DECISION AIDING VIA INTELLIGENT INTERFACES

Decision aiding has progressed along several lines over the past two decades. The first wave saw simple computational devices capable of calculating optimal routes, attrition rates, and event probabilities. The second wave attempted to exploit technology—such as expert systems, interactive, direct manipulation graphics, and map-based interaction routines. The third wave is more difficult to define, save for selected emphases on requirements engineering (and larger decision systems engineering issues), analytical methods, and specific user–computer interaction tools and techniques.

It can be argued that the user–computer interface represents the computer-based decision-making "cockpit," and that it is now possible to satisfy complex, changing requirements via decision aids that are domain and user smart and situationally responsive. In other words:

- Many current decision aids and support systems were not conceived via an explicit allocation filter, that is, their designers have not examined carefully enough opportunities for shifting tasks from their operators to their systems.

- Information and computational technology has progressed far enough to support automation across a large class of decision-making activities, activities that now result in high operator-system workload.

- Analytical methodology—broadly defined—has not been optimally used to model decision-aiding requirements or to support operational functions and tasks; there have been far too few attempts to develop multiple methods-based decision-aiding systems; systems that because of their multiple methods architectures, could relieve operators of many current input, processing, and output inspection tasks.

- Although user–computer interface technology has begun to receive serious attention from the systems analysis, design, research, and development community, its potential for adaptive aiding has not been realized.

- We have not shifted interaction "paradigms"; we still largely believe that the computer-supported decision-making process requires human operators to frame problems, input data, construct queries, interpret map displays, directly manipulate displays, and "send" data and information to other similarly overloaded operators.

Adaptive Task Allocation

There are any number of domains that do not currently enjoy optimal human-system task allocation. There are countless systems intended to support complex inductive problem solving, and many simple deductive domains that beg for software-intensive solutions. There are overstressed human weapons directors working without the kind of intelligent system support well within the state of the art. There are some relatively straightforward steps that can be taken to allocate optimally tasks across human and electronic or software-intensive system components. We need insight into the allocation process and techniques to distribute tasks across users and the systems that support them. One way is to examine the characteristics of the tasks that must be supported by the system.

Task Characteristics

There are several ways to categorize problems. Some approaches deal with simple criteria, such as the extent to which a problem is time-sensitive or analytically complex. Shneiderman (1992) developed an instructive list:

Humans Better	*Machines Better*
Sense low-level stimuli	Sense stimuli outside human's range
Detect stimuli in noisy background	Count or measure physical quantities
Recognize constant patterns in varying situations	Store quantities of coded information
Sense unusual and unexpected events	Monitor prespecified events
Remember principles and strategies	Recall quantities of detailed information
Retrieve details without a priori connection	Process quantitative data in prespecified ways
Draw on experience and adapt decisions to situations	
Select alternatives if original approach fails	
Reason inductively: generalize from observations	Reason deductively: infer from a principle
Act in unanticipated emergencies and novel situations	Perform repetitive preprogrammed actions
Apply principles to solve varied problems	
Make subjective evaluations	Perform several activities simultaneously
Develop new solutions	
Concentrate on important tasks when overload occurs	Maintain operations under heavy load
Adapt physical response to changes in situation	Maintain performance over extended periods of time

Another way to categorize problems locates them within multidimensional spaces, such as the one developed by Hermann (1969). Hermann's "situational cube" locates problems along three dimensions: threat, decision time, and awareness. A crisis, for example, simply because of time constraints, calls for more system autonomy and less human deliberation than so-called "routine" situations. As we attempt to design next generation decision aids and support systems, we must look to the situational milieu for guidance about optimal task allocation. Next generation "super-cockpits," for example, call for enormous standoff and onboard processing power. But when the cockpit moves from the commercial airways to the battlefield, task allocation is exacerbated by short-time, high-threat, or surprise situations, in which combat pilots operate under ultra-real-time conditions. When an adversary surface-to-air missile has locked onto an aircraft and the missile has been launched, the system will clearly respond immediately and, on selected occasions, override any pilot commands that might endanger the pilot, the aircraft, and/or the mission.

Optimal Task Allocation

The heuristics for optimal task allocation can be gleaned from experience, empirical data from successful systems, and research on inherent human and electronic computational capabilities. It is indeed possible to allocate tasks given sets of problem characteristics and situational constraints. Another key aspect of the allocation process is operator-system interaction technology or the means by which information and decisions are conferenced and communicated to operators, as appropriate. Advanced display technology can facilitate the rapid communication of system responses to stimuli too numerous and complex for humans to process. Input technologies such as direct manipulation interfaces and speech can facilitate rapid operator action and reaction.

Situational constraints will predict to allocation strategies which, in turn, will predict to analytical methods. But we need a strategy that is based on empirical evidence, validated research about human and electronic computational capabilities, and experience. When this research is synthesized, heuristics for optimal task allocation will emerge.

Adaptive User–Computer Interaction

User–computer interaction (UCI) is emerging as one of the key information technologies of the 1990s. Why? Because the interface *is* the system to many users. As such it represents an opportunity to organize, orchestrate, assess, and prescribe domain behavior. In other words, it represents an opportunity to change the way we interact with interactive systems.

The May 24, 1993, issue of *MacWeek*, for example, talks about intelligent "agents" for enhanced user–computer interaction. Agents, it is argued, can observe interactive behavior and (via a set of very simple heuristics) suggest interaction strategies that might save users lots of time and effort—especially in high workload situations. The concept is by no means new, but the commercialization of the technology in cost-effective software "accessories" is.

Adaptation can be stimulated by several aspects of the interaction process. The first is the character of the situation at hand; the second is the nature of current user-specific interaction; whereas the third is the system's models of the user and the domain.

UCI technology is also progressing to the point where it is possible to communicate in multiple ways to multiple users for different purposes. We have progressed far beyond simple color or graphics-based "feedback": Systems are now capable of watching, understanding, and even "learning" (not in the same sense that a child learns how to tie a shoe and then transfers that knowledge to tying down a tarp, but impressive nonetheless).

Adaptive Analytical Methodology

Computer-based problem solving involves—at a minimum—a set of processes that monitor an environment, assess a situation, generate options, implement options, and assess impact (with the cycle beginning again, and again, and so on).

There is an opportunity to introduce and demonstrate a new approach to aiding based on the use of multiple methods and models. Given today's computing technology and the state of our analytical methodology, it is possible to design systems that conduct analyses from multiple analytical perspectives. We should redefine the tasks-, or methods-matching process to include the integration of multiple methods and models to increase "hits" and decrease "false alarms."

Multiple model-based decision aiding is a new concept directly pertinent to reliability and accuracy. There are a variety of monitoring, diagnosis, assessment, and warning methods available. The challenge lies in the extent to which we can identify and integrate the "right" (multiple) methods into the domains we select to aid.

The design process appears next. The basic concept assumes that monitoring, diagnosis, and situation assessment tasks can be addressed by multiple qualitative and quantitative models and methods, and that, when combined with advanced user–computer interface technology, can yield prototypes:

Monitoring, Diagnosis, Situation Assessment & "Warning"	→	Multiple (Quantitative & Qualitative) Models & Analytical Methods	→	Advanced Information & User Computer Interface Technology	→	Prototyping

The methods architecture calls for multiple methods working in parallel. These would include a suite of tools and techniques that crosscut the epistemological spectrum. Procedural preprocessing would perhaps involve the use of neural networks, fuzzy logic systems, genetic learning algorithms, conventional (heuristic) expert systems, multivariate models, statistical pattern recognition, and a variety of other probabilistic methods, tools, and techniques, such as first-order Bayesian mechanisms and Markov chains. The key is the allocation of computational resources at the front end of the tactical monitoring process. Qualitative and quantitative models, tools, and techniques can be used to filter incoming sensor data; in effect, a processing "race" should be set up to distill the data to its most diagnostic properties as quickly as possible.

The entire architecture assumes that if a match is made relatively early in the monitoring process, then systems can, in turn, proceed more confidently toward the warning and management phases. In other words, the system can shift its methodological orientation from probabilistic methods, tools, and techniques to those more inclined toward optimization.

The front-end architecture presumes large data and knowledge bases. Data, information, and knowledge about alternative emitter patterns, weapons signatures and capabilities, aircraft capabilities, deceptive patterns, and the like must be stored in mission contexts, that is, as part of the mission planning and execution process. In a manner of speaking, we are talking here about the incarnation of intelligent scenarios embedded in the pre- and postprocessing methods modules.

Another way of conceiving of the architecture is to understand front-end processes as data-driven, and subsequent stages as parameter-driven. Front-end methods can range from data-driven to judgment-driven to heuristic-driven, as well as parameter-driven methods, where appropriate.

The selection of specific parameter-, data-, judgment-, and heuristic-based methods will depend on the nature of available data and information and the validity of existing or potential knowledge bases.

Some Recommendations

All of this leads inevitably to a set of prescriptions about how to design next generation decision aids and larger support systems:

- Domain and associated tasks analyses can be used to identify functions, tasks, and processes that can be automated; such analyses can identify the functions, tasks, and processes that can be targeted for operator/system allocation (or reallocation in the case of decision aid reengineering).
- Situational variables can help with the allocation/reallocation process.
- Information technology can support task allocation (reallocation); it can also constrain it.
- Doctrinal and professional "cultural" considerations should be examined prior to any allocation or reallocation activities.
- Constraints should be assessed during the allocation/reallocation process.

- The user–computer interface (UCI) can serve as a window to quasi- and fully automated systems activities.

These and related prescriptions suggest that there are countless unexploited opportunities for automation, opportunities that technology has placed well within our reach. Some examples include:

- Systems that can assess situations and make recommendations.
- Systems that can adaptively calculate target values and automatically allocate fire.
- Systems that can automatically monitor, assess, and warn of impeding events and conditions.
- Systems that can adaptively allocate engagement resources, such as electronic warfare resources.
- Systems that can monitor the effectiveness of individual operators and operations teams, calibrate strengths and weaknesses, and recommend alternative staffing solutions.
- Systems that can automatically control selected weapons systems.
- Systems that can maneuver platforms, engage, and reconfigure.
- Systems that can communicate articulately with operators; operators can communicate with their systems.
- Systems that can adapt to changing situations.
- Systems that can generate and evaluate options.
- Systems that can adapt to situations, operators, feedback, and other stimuli in real-time and ultra-real-time.
- Systems that can invoke indicator-based I&W processes that can yield probabilistic warnings.
- Systems that can display data, information, and knowledge necessary to understand and communicate "tactical" and "strategic"situations, options, and objectives.
- Systems that can receive and process data, identify and route messages, and update situational profiles.

These and related activities can be achieved via the creative mix of requirements data, technology, and implementation constraints. Perhaps larger questions involve the desirability of the capabilities. There are powerful "cultural" forces that help define the acceptable range of acceptable automation. There are vested interests at stake as well. Perhaps the best we can achieve is incremental.

In an environment where "technology push" frequently overwhelms "requirements pull," the caution surrounding the movement toward automation has impeded technological possibilities, while simultaneously leaving many requirements unfulfilled. This stalemate has resulted in the underapplication of technology to unsolved "manual" and unnecessarily interactive problems.

Just as corporate America has discovered that it does not need anywhere near the number of middle managers it hired in the 1970s and 1980s, systems designers, managers, and sponsors will recognize the inherent inefficiency of much human–computer interaction.

The next 5 to 10 years will see pressure from decreasing budgets, less and less personnel, evolving doctrine, and opportunities from information technology and user–computer interface technology revolutionize the computer-based problem-solving process. What today we regard as within the exclusive preserve of human decision makers, tomorrow we will regard as fair game for automated methods, tools, and techniques. The interface will stand between decision makers and their computer-based assistants and—at times—full-fledged colleagues; it will also at times ask for help, simply state the obvious, and—once in a while—take complete charge of situations unfolding at rates that far exceed human capacities.

STANDARDS

Not very long ago all kinds of questions about interface "look and feel" perplexed the designers of user–computer interfaces. But since the development of the Xerox Star system, the Apple Lisa, the Macintosh, and, much more recently, Microsoft Windows, a huge percentage of questions about "optimal" interaction features has been answered. Although this is not necessarily to imply that all of the features of the modern interface (Macintosh/Windows/X Window/OSF/Motif) will inevitably lead to enhanced performance, it is to suggest that we have today a de facto UCI standard. The interface guidelines published by the "owners" of operating systems now require application program vendors to design and develop software according to UCI standards.

The menu structures, interaction structures (direct manipulation, for example), and other features by and large define the overall interaction process. Consistency is the watchword here—and it is a sensible one. At the same time, there is still a lot of action within the menu structure, with the ultimate window into which we place data, queries, and interactive displays. It is here where creative UCI designers will work.

INFORMATION TECHNOLOGY

Multimedia, virtual reality, three-dimensional displays, teleconferencing, computer-supported cooperative work, and so many other technologies are emerging. How will they figure in the design and development of user–computer interfaces? Research and development is necessary to determine the situations, tasks, and domains where emerging technologies can best be utilized. Systems engineers also need to assess the technology's cost-effectiveness, maintenance requirements, and role in configuration management.

THE NEED FOR EVALUATION

All of the ideas in this book and trends, issues, and recommendations in this chapter require insight gleaned from testing and evaluation. Technology is emerging as fast as we can assess its impact—or perhaps faster! We must keep the evaluation pace up so we can develop guidelines for cost-effective UCI design and development. Qualitative and quantitative methods, tools, and techniques must be continually improved so assessments about technologies can be made quickly and diagnostically.

A NET ASSESSMENT

The field of UCI design and development is healthy. Over the past decade UCI consciousness has been raised by several orders of magnitude. Even the most unsympathetic systems designers and developers have come to appreciate the importance of the interface and the relationship between the nature and operation of the interface and productivity.

This book has attempted to shed some light on one aspect of the design and development process. It represents a small—but hopefully significant— step in the overall UCI design and development process. But it constitutes only a small step in the march toward fully adaptive, intelligent, cognitive friendly interfaces. Just as the past decade has proven monumental in the design and development of user-computer interfaces, the next decade will introduce new UCI technologies, design philosophies, and case studies. As we approach the 21st century, we can expect the form and content of the user–computer interaction process to evolve creatively and practically.

References

Adams, D. R. (1990). *Hypercard and Macintosh: A primer.* New York: McGraw-Hill.

Adelman, L. (1981). The influence of formal, substantive, and contextual task properties on the relative effectiveness of different forms of feedback in multiple-cue probability learning tasks. *Organizational Behavior and Human Performance, 27,* 423–442.

Adelman, L. (1984). Real-time computer support for decision analysis in a group setting: Another class of decision support systems. *Interfaces, 14,* 75–83.

Adelman, L. (1988). Separation of facts and values. In B. Brehmer & C. R. B. Joyce (Eds.), *Human judgment: The SJT view.* Amsterdam: North-Holland.

Adelman, L. (1991). Experiments, quasi-experiments, and case studies: A review of empirical methods for evaluating decision support systems. *IEEE Transactions on Systems, Man, and Cybernetics, 21*(2), 293–301.

Adelman, L. (1992). *Evaluating decision support and expert systems.* New York: John Wiley & Sons.

Adelman, L., Black, P., Marvin, F., & Sak, S. (1992). *Information order effects on expert judgment* (Tech. Rep. 92-3). Fairfax, VA: Decision Science Consortium, Inc.

Adelman, L., & Bresnick, T. A. (1992). Examining the effect of information sequence on expert judgment: An experiment with Patriot air defense officers using the Patriot air defense simulator. *Organizational Behavior and Human Decision Processes, 53,* 204–228.

Adelman, L., Cohen, M. S., Bresnick, T. A., Chinnis, Jr., J. O., & Laskey, K. B. (1993). Real-time expert system interfaces, cognitive processes, and task performance: An empirical assessment. *Human Factors, 35,* 243–261.

Adelman, L., Gualtieri, J., & Stanford, S. (in press). Toward understanding the option generation process: An experiment using protocol analysis. *Organizational Behavior and Human Decision Processes.*

Adelman, L., Stewart, T. R., & Hammond, K. R. (1975). A case history of the application of social judgment theory to policy formation. *Policy Sciences, 6,* 137–159.

Adelman, L., Tolcott, M. A., & Bresnick, T. A. (1993). Examining the effect of information order on expert judgment. *Organizational Behavior and Human Decision Processes, 56,* 348–369.

Ahituv, N., Even-Tsur, D., & Sadan, B. (1986). Procedures and practices for conducting postevaluation of information systems. *Journal of Information Systems Management, 3*(2).

241

Aiken, P. H. (1989). *A hypermedia workstation for software engineering*. Unpublished doctoral dissertation, George Mason University, Fairfax, VA.

Ambron, R., & Hooper, C. (1987). *Interactive multimedia: Visions of multimedia for developers, educators and information providers*. Redmond, WA: Microsoft Publishers.

Ameen, D. A. (1989). Systems performance evaluation. *Journal of Systems Management, 40*(3).

Anderson, J. R. (1982). Acquisition of cognitive skill. *Psychological Review, 89*, 369–406.

Anderson, J. R. (1983). *The architecture of cognition*. Cambridge, MA: Harvard University Press.

Andriole, S. J. (1983). *Interactive computer-based systems*. New York: Petrocelli Books.

Andriole, S. J. (1988). Storyboard prototyping for requirements verification. *Large Scale Systems, 12*, 231–247.

Andriole, S. J. (1989). *Handbook for the design, development, evaluation, and application of interactive decision support systems*. Princeton, NJ: Petrocelli.

Andriole, S. J. (1990). *Modern life cycling: Some system design principles for the 90s*. Fairfax, VA: AFCEA International Press.

Andriole, S. J. (1992). *Rapid application prototyping*. Wellesley, MA: QED Information Sciences.

Andriole, S. J., & Adelman, L. (1989, October). Cognitive engineering of advanced information technology for Air Force systems design and deployment (AD-A218 558). International Information Systems.

Andriole, S. J., & Adelman, L. (1991). Prospects for cognitive systems engineering. *Proceedings of 1991 IEEE International Conference on Systems, Man, and Cybernetics* (pp. 743–747).

Andriole, S. J., Ehrhart, L. S., Aiken, P. H., & Matyskiela, W. M. (1989, June). Group decision support system prototypes for army theater planning and counter-terrorism crisis management. *Proceedings of the Third Annual Symposium on Command and Control Research*, Washington, DC.

Andriole, S. J., & Freeman, P. W. (1993). Software systems engineering: The case for a new discipline. *Software Engineering Journal, 4*(3), 455–468.

Ashton, A. H., & Ashton, R. H. (1988). Sequential belief revision in auditing. *The Accounting Review, 63*, 623–641.

Ashton, R. H., & Ashton, A. H. (1990). Evidence-responsiveness in professional judgment: Effects of positive versus negative evidence and presentation mode. *Organizational Behavior and Human Decision Processes, 46*, 1–19.

Baecker, R., & Buxton, W. (1987). *Readings in human-computer interaction*. Los Altos, CA: Morgan Kaufman Publishers.

Bailey, J. E., & Pearson, S. (1983). Development of a tool for measuring and analyzing computer user satisfaction. *Management Science, 29*, 530–545.

Bailey, R. W. (1989). *Human performance engineering: Using human factors/ergonomics to achieve computer system usability*. Englewood Cliffs, NJ: Prentice-Hall.

Balke, W. M., Hammond, K. R., & Meyer, G. D. (1972). An alternative approach to labor-management relations. *Administrative Science Quarterly, 18*, 311–327.

Balzer, W. K., Doherty, M. E., & O'Connor, R. (1989). Effects of cognitive feedback on performance. *Psychological Bulletin, 106*, 410–433.

Bartlett, F. C. (1983). *Remembering*. Cambridge, England: Cambridge University Press.

Bazerman, M. H. (1990). *Judgment in managerial decision making* (2nd ed.). New York: John Wiley & Sons.

Beach, L. R. (1990). *Image theory: Decision making in personal and organizational contexts*. New York: John Wiley & Sons.

Beach, L. R., & Mitchell, T. R. (1987). Image theory: Principles, goals, and plans in decision making. *Acta Psychologica, 66*, 201–220.

Beach, T. (1973, May). Color in paperwork. *The Office*, pp. 80–86.

Benbasat, I., & Taylor, R. N. (1982). Behavioral aspects of information processing for the design of management information systems. *IEEE Transactions on Systems, Man, and Cybernetics, SMC-12*, 439–450.

Berlyne, D. (1960). *Conflict, arousal, and curiosity.* New York: McGraw-Hill.

Bice, K., & Lewis, C. (1989). *Wings for the mind: Conference Proceedings: Computer Human Interaction.* Reading, MA: Addison-Wesley Publishing.

Bishop, Y. M. M. (1990). Behavioral concerns in support system design. In A. P. Sage (Ed.), *Concise encyclopedia of information processing in systems and organizations* (pp. 449–454). Oxford: Pergamon Press.

Bishop, Y. M. M., Fienberg, S. E., & Holland, P. W. (1975). *Discrete multivariate analysis: Theory and practice.* Cambridge, MA: MIT Press.

Blanchard, B. (1990). *Systems engineering.* Englewood Cliffs, NJ: Prentice Hall.

Blattberg, R. C., & Hoch, S. J. (1990). Database models and managerial intuition: 50% model + 50% manager. *Management Science, 36,* 887–899.

Bobrow, D. G., & Norman, D. A. (1975). Some principles of memory schemata. In D. G. Bobrow & A. Collins (Eds.), *Representation and understanding: Studies in cognitive science.* New York: Academic Press.

Bolt, R. A. (1984). *The human interface: Where people and computers meet.* Belmont, CA: Lifetime Learning Publications.

Bouwman, M. J. (1984). Expert vs. novice decision making in accounting: A summary. *Accounting Organizations and Society, 9,* 325–327.

Breen, P. J., Miller-Jacobs, P. E., & Miller-Jacobs, H. (1987). Color displays applied to command, control, and communication systems. In H. J. Durrett (Ed.), *Color and the computer.* FL: Academic Press.

Brehmer, B. (1986). Social judgment theory and the analysis of interpersonal conflict. *Psychological Bulletin, 83,* 985–1003.

Brehmer, B., & Joyce, C. R. B. (Eds.). (1988). *Human judgment: The SJT view.* Amsterdam: North-Holland.

Brehmer, B., Kuylenstierna, J., & Liljergen, J. E. (1974). Effect of function form and cue validity on the subjects' hypotheses in probabilistic inference tasks. *Organizational Behavior and Human Performance, 11,* 338–354.

Brown, R. V., Kahr, A. S., & Peterson, C. R. (1974). *Decision analysis for the manager.* New York: Holt, Rinehart & Winston.

Byers, J. C., Bittner, A. C., & Hill, S. G. (1989). Traditional and raw Task Load Index (TLX) correlations: Are paired comparisons necessary? In A. Mital (Ed.), *Advances in industrial ergonomics and safety* (pp. 481–485). London: Taylor & Francis.

Byers, J. C., Bittner, A. C., Hill, S. G., Zaklad, A. L., & Christ, R. E. (1988). Workload assessment of a remotely piloted vehicle system. *Proceedings of the Human Factors Society 32nd Annual Meeting.* Santa Monica, CA: Human Factors Society.

Camerer, C. F., & Johnson, E. J. (1991). The process-performance paradox in expert judgment: How can experts know so much and predict so badly? In K. A. Ericsson & J. Smith (Eds.), *Toward a general theory of expertise: Prospects and limits.* New York: Cambridge University Press.

Campbell, D. T., & Stanley, J. C. (1969). *Experimental and quasi-experimental designs for research.* Chicago: McNally & Company.

Card, S. K., Moran, T. P., & Newell, A. (1983). *The psychology of human-computer interaction.* Hillsdale, NJ: Lawrence Erlbaum Associates.

Carroll, J. M. (1984). Mental models and software human factors: An overview. (Research Rep. RC 10616 [#47016]). Yorktown Heights, NY: IBM Watson Research Center.

Carroll, J. M. (Ed.). (1987). *Interfacing thought: Cognitive aspects of human–computer interaction.* Cambridge, MA: MIT Press.

Carroll, J. M. (1989). Taking artifacts seriously. In S. Maass & H. Oberquelle (Eds.), *Software ergonomie' 89: Aufgabenorientierte systemgestaltung und functionalitaet.* Stuttgart: Teubner.

Carroll, J. M., & Thomas, J. C. (1982). Metaphor and the cognitive representation of computing systems. *IEEE Transactions on Systems, Man, and Cybernetics, SMC-12*, 107–116.

Carter, J. A., Jr. (1986). A taxonomy of user-oriented functions. *International Journal of Man-Machine Studies, 24*, 195–292.

Chase, W. G., & Ericsson, K. A. (1981). Skilled memory. In J. R. Anderson (Ed.), *Cognitive skills and their acquisition*. Hillsdale, NJ: Lawrence Erlbaum Associates.

Chase, W. G., & Ericsson, K. A. (1982). Skill and working memory. In G. H. Bower (Ed.), *The psychology of learning and motivation: Advances in research and theory* (Vol. 16). New York: Academic Press.

Chase, W. G., & Simon, H. A. (1973). Perception in chess. *Cognitive Psychology, 4*, 55–81.

Chignell, M. H. (1990). A taxonomy of user interface terminology. *SIGCHI Bulletin, 21*(4), 27–34.

Chinnis, J. O., Jr., Cohen, M. S., & Bresnick, T. A. (1985). *Human and computer task allocation in air defense systems* (Army Research Institute Tech. Rep. 691). Alexandria, VA: Army Research Institute.

Coffey, J. L. (1961). A comparison of vertical and horizontal arrangements of alphanumeric material—Experiment I. *Human Factors, 3*, 93–98.

Cohen, M. S. (1987). *When the worst case is best: Mental models, uncertainty, and decision aids. Impact and potential of decision research on decision aiding* (Report of a Department of Defense Roundtable Workshop). Washington, DC: American Psychological Association.

Cohen, M. S. (1991). *Situation assessment skills in battlefield planning*. Arlington, VA: Cognitive Technologies, Inc.

Cohen, M. S., Adelman, L., Bresnick, T. A., Chinnis, J. O., Jr., & Laskey, K. B. (1988). *Human and computer task allocation in air defense systems: Final report*. (Tech. Rep. 87-16). Reston, VA: Decision Science Consortium, Inc.

Cohen, M. S., & Hull, K. (1990). *Recognition and metacognition in sonar classification*. (Draft Tech. Rep.). Reston, VA: Decision Science Consortium, Inc.

Connolly, T., & Wagner, W. G. (1988). Decision cycles. In R. L. Cardy, S. M. Puffer, & M. M. Newman (Eds.), *Advances in information processing in organizations* (Vol. 3, pp. 183–205). Greenwich, CT: JAI Press.

Conover, W. J. (1980). *Practical nonparametric statistics* (2nd ed.). New York: John Wiley & Sons.

Cook, R. L., & Stewart, T. R. (1975). A comparison of seven methods for obtaining subjective descriptions of judgmental policy. *Organizational Behavior and Human Performance, 13*, 31–45.

Craik, K. J. W. (1943). *The nature of explanation*. Cambridge: Cambridge University Press.

Daft, R. L., & Lengel, R. H. (1986). Organizational information requirements, media richness, and structural designs. *Management Science, 32*, 554–571.

Dawes, R. M. (1979). The robust beauty of improper linear models. *American Psychologist, 34*, 571–582.

Dawes, R. M., & Corrigan, B. (1974). Linear models in decision making. *Psychological Bulletin, 81*, 95–106.

Dearborn, D. C., & Simon, H. A. (1958). Selective perception: A note on the departmental identification of executives. *Sociometry, 21*, 140–144.

Defense Advanced Research Projects Agency (DARPA)/Massachusetts Institute of Technology (MIT). (1988). *DARPA neural network study*. Fairfax, VA: AFCEA International Press.

DeSanctis, G., & Gallupe, R. B. (1987). A foundation for the study of group decision support systems. *Management Science, 33*, 589–609.

Dixon, W. J. (Ed.). (1977). *Biomedical computer programs P-series*. Los Angeles, CA: University of California Press.

Dodd, D. H., & White, R. M., Jr. (1980). *Cognition: Mental structures and processes*. Boston, MA: Allyn & Bacon.

Dreyfuss, H. (1955). *Designing for people.* New York: Simon & Schuster.

Durrett, H. J. (Ed.). (1987). *Color and the computer.* Orlando, FL: Academic Press.

Ebert, R. J., & Kruse, T. E. (1978). Bootstrapping the security analyst. *Journal of Applied Psychology, 63,* 110–119.

Edwards, W. (1968). Conservatism in human information processing. In B. Kleinmuntz (Ed.). *Formal representation of human judgment.* New York: John Wiley & Sons.

Ehrhart, L. S. (1990). *Cognitive systems engineering: New directions for the design of the human-computer interface in database management systems.* Paper presented at the College of Systems Engineering, George Mason University, Fairfax, VA.

Einhorn, H. J., & Hogarth, R. M. (1975). Unit weighting schemes for decision making. *Organizational Behavior and Human Performance, 13,* 171–192.

Einhorn, H. J., & Hogarth, R. M. (1978). Confidence in judgment: Persistence of the illusion of validity. *Psychological Review, 85,* 395–416.

Einhorn, H. J., & Hogarth, R. M. (1981). Behavioral decision theory: Processes of judgment and choice. *Annual Review of Psychology, 32,* 53–88.

Einhorn, H. J., & Hogarth, R. M. (1985). Ambiguity and uncertainty in probabilistic inference. *Psychological Review, 92,* 433–461.

Einhorn, H. J., & Hogarth, R. M. (1986). Judging probable cause. *Psychological Bulletin, 99,* 3–19.

Einhorn, H. J., & Hogarth, R. M. (1987). *Adaptation and inertia in belief updating: The contrast-inertia model.* (Tech. Rep., Center for Decision Research). Chicago: University of Chicago.

Einhorn, H. J., Hogarth, R. M., & Smith, J. (1991). Prospects and limits of the empirical study of expertise: An introduction. In K. A. Ericsson & H. A. Simon. (1984). *Protocol analysis: Verbal reports as data.* Cambridge, MA: MIT Press.

Eisner, H. E. (1989). *Computer-aided systems engineering.* Englewood Cliffs, NJ: Prentice-Hall.

Endsley, M. R. (1988). Design and evaluation for situational awareness. *Proceedings of the Human Factors Society 32nd Annual Meeting.* Santa Monica, CA: Human Factors Society.

Ericsson, K. A., & Smith, J. (Eds.). (   ).*Toward a general theory of expertise: Prospects and limits.* New York: Cambridge University Press.

Fairchild, K. M., Meredith, L. G., & Wexelblat, A. D. (1988a). *A formal structure for automatic icons* [MCC Tech. Rep. No. STP-311-88].

Fairchild, K. M., Meredith, L. G., & Wexelblat, A. D. (1988b). *The tourist artificial reality* [MCC Tech. Rep. No. STP-310-88].

Falzon, P. (1990). *Cognitive ergonomics: Understanding, learning, and designing human-computer interaction.* New York: Academic Press.

Fischhoff, B. (1975). Hindsight = foresight: The effect of outcome knowledge on judgment under uncertainty. *Journal of Experimental Psychology: Human Perception and Performance, 1,* 288–299.

Fischhoff, B., Slovic, P., & Lichtenstein, S. (1978). Fault trees: Sensitivity of estimated failure probabilities to problem representation. *Journal of Experimental Psychology: Human Perception and Performance, 4,* 330–344.

Fischhoff, B., Slovic, P., & Lichtenstein, S. (1980). Knowing what you want: Measuring labile values. In T. Wallsten (Ed.), *Cognitive processes in choice and decision behavior* (pp. 64–85). Hillsdale, NJ: Lawrence Erlbaum Associates.

Fitzgerald, J. R., & Grossman, J. D. (1987). Decision aids—Who needs them? In M. A. Tolcott & V. E. Holt (Eds.), *Impact and potential of decision research on decision aiding: Report of a department of defense roundtable workshop.* Washington, DC: American Psychological Association.

Flanagan, J. C. (1954). The critical incident technique. *Psychological Bulletin, 51,* 327–358.

Fracker, M. L. (1988). A theory of situation assessment: Implications for measuring situation awareness. *Proceedings of the Human Factors Society 32rd Annual Meeting.* Santa Monica, CA: Human Factors Society.

Gardner-Bonneau, D. J. (1993, April). [Comment from the editor.] *Ergonomics in design.*

Glaser, R., & Chi, M. T. H. (1988). Overview. In M. T. H. Chi, R. Glaser, & M. J. Farr (Eds.), *The nature of expertise.* Hillsdale, NJ: Lawrence Erlbaum Associates.

Goldberg, L. R. (1970). Man versus model of man: A rationale, plus some evidence for a method of improving on clinical inferences. *Psychological Bulletin, 73,* 422–432.

Gould, J. D., & Lewis, C. (1985). Designing for usability: Key principles and what designers think. *Communications of the ACM, 28,* 300–311.

de Groot, A. D. (1965). *Thought and choice in chess.* The Hague: Mouton.

de Groot, A. D. (1966). Perception and memory versus thinking. In B. Kleinmuntz (Ed.), *Problem solving.* New York: John Wiley & Sons.

Hamel, K. (1988, October). *Construct measurement in management information systems research: The case of user satisfaction and involvement.* Paper presented at the 26th Annual Joint National Meeting of the Operations Research Society of America/The Institute of Management Sciences, Denver, CO.

Hammond, K. R. (1966). Probabilistic functionalism: Egon Brunswik's integration of the history, theory, and method of psychology. In K. R. Hammond (Ed.), *The psychology of Egon Brunswik.* New York: Holt, Rinehart & Winston.

Hammond, K. R. (1971). Computer graphics as an aid to learning. *Science, 172,* 903–908.

Hammond, K. R. (1976). *Facilitation of interpersonal learning and conflict reduction by on-line communication.* Boulder, CO: University of Colorado.

Hammond, K. R. (1988). Judgment and decision making in dynamic tasks. *Information and Decision Technologies, 14,* 3–14.

Hammond, K. R., & Adelman, L. (1976). Science, values, and human judgment. *Science, 194,* 389–396.

Hammond, K. R., & Boyle, P. J. R. (1971). Quasi-rationality, quarrels and new conceptions of feedback. *Bulletin British Psychological Society, 24,* 103–113.

Hammond, K. R., & Brehmer, B. (1973). Quasi-rationality and distrust: Implications for international conflict. In L. Rappoport & D. Summers (Eds.), *Human judgment and social interaction.* New York: Holt, Rinehart & Winston.

Hammond, K. R., & Grassia, J. (1985). The cognitive side of conflict: From theory to resolution of policy disputes. In S. Oskamp (Ed.), *Applied Social Psychology Annual: Vol. 6. International Conflict and National Public Policy Issues* (pp. 233–254). Beverly Hills: Sage.

Hammond, K. R., Hamm, R. M., Grassia, J., & Pearson, T. (1987). Direct comparison of the efficacy of intuitive and analytical cognition in expert judgment. *IEEE Transactions in Systems, Man, and Cybernetics, SMC-17,* 753–770.

Hammond, K. R., Hursch, C. J., & Todd, F. J. (1964). Analyzing the components of clinical judgment. *Psychological Review, 71,* 438–456.

Hammond, K. R., Rohrbaugh, J., Mumpower, J., & Adelman, L. (1977). Social judgment theory: Applications in policy formation. In M. F. Kaplan & S. Schwartz (Eds.), *Human judgement & decision processes in applied settings.* New York: Academic Press.

Hammond, K. R., Stewart, T. R., Brehmer, B., & Steinmann, D. O. (1975). Social judgment theory. In M. F. Kaplan & S. Schwartz (Eds.), *Human judgment and decision processes* (pp. 271–312). New York: Academic Press.

Hammond, K. R., & Summers, D. A. (1972). Cognitive control. *Psychological Review, 79,* 58–67.

Hart, S. G. (1987). Background, description, and application of the NASA TLX. *Proceedings of the Department of Defense Human Factors Engineering Technical Advisory Group, Dayton, Ohio.*

Hart, S. G. (1988). Development of NASA-TLX: Results of empirical and theoretical research. In P. A. Hancock & N. Meshkati (Eds.), *Human Mental Workload* (pp. 239–250). Amsterdam: North-Holland.

Hartman, B. O. (1961). Time and load factors in astronaut proficiency. In B. E. Flatery (Ed.), *Symposium on Psychophysiological Aspects of Space Flight.* New York: Columbia University Press.

Hartson, H. R., & Hix, D. (1989). Human–computer interface development: Concepts and systems for its management. *ACM Computing Surveys, 21,* 5–91.

Hecht-Nielson, R. (1988, April). Neural networks. *IEEE Spectrum,* pp. 352–358.

Helander, M. (Ed.). (1988). *Handbook of human–computer interaction.* Amsterdam: North-Holland.

Hermann, C. F. (1969). International crisis as a situational variable. In C. F. Hermann (Ed.), *International crises: Insights from behavioral research.* Glencoe, IL: The Free Press.

Hill, S. G., Iavecchia, H. P., Byers, J. C., Bittner, A. C., Zaklad, A. L., & Christ, R. E. (1992). Comparison of four subjective workload rating scales. *Human Factors, 34,* 429–439.

Hirschheim, R. A. (1983). Assessing participative systems design: Some conclusions from an exploratory study. *Information and Management, 6,* 317–327.

Hirschheim, R. A. (1985). User experience with and assessment of participative systems design. *MIS Quarterly, 12,* 295–303.

Hoadley, E. D. (1990). Investigating the effects of color. *Communications of the ACM, 33,* 120–125.

Hogarth, R. M. (1975). Cognitive processes and the assessment of subjective probability distributions. *Journal of the American Statistical Association, 70,* 271–289.

Hogarth, R. M. (1978). A note on aggregating opinions. *Organizational Behavior & Human Performance, 21,* 40–46.

Hogarth, R. M. (1981). Beyond discrete biases: Functional & dysfunctional aspects of judgmental heuristics. *Psychological Bulletin, 90,* 197–217.

Hogarth, R. M. (1987). *Judgment and choice* (2nd ed.). New York: Wiley-Interscience.

Hogarth, R. M., & Einhorn, H. J. (1992). Order effects in belief updating: The belief-adjustment model. *Cognitive Psychology, 24,* 1–55.

Holland, J. H., Holyoak, K. J., Nisbett, R. E., & Thagard, P. R. (1986). *Induction: Processes of inference, learning, and discovery.* Cambridge, MA: MIT Press.

Hollnagel, E., & Woods, D. D. (1983). Cognitive systems engineering: New wine in new bottles. *International Journal of Man-Machine Studies, 18,* 583–600.

Holyoak, K. (1991). Symbolic connectionism: Toward third-generation theories of expertise. In K. A. Ericsson & J. Smith (Eds.), *Toward a general theory of expertise: Prospects and limits.* New York: Cambridge University Press.

Hopple, G. W. (1986). Decision aiding dangers: The law of the hammer and other maxims. *IEEE Transactions on Systems, Man and Cybernetics, SMC-16(6),* 230–239.

Howell, W. C., & Tate, J. D. (1966). Influence of display, response, and response set factors upon the storage of spatial information in complex displays. *Journal of Applied Psychology, 50,* 73–80.

Huber, G. P. (1980). *Managerial decision making.* Glenview, IL: Scott, Foresman, & Company.

Isenberg, D. J. (1986). Thinking and managing: A verbal protocol analysis of managerial problem solving. *Academy of Management Journal, 29,* 775–788.

Ives, B., Olson, M., & Baroudi, J. (1983). The measurement of user information satisfaction. *Communications of the ACM, 26(10),* 54–65.

Jacoby, J. (1987). Consumer research: A state of art review. *Journal of Marketing, 42(2),* 87–96.

Janis, I. L. (1972). *Victims of groupthink.* Boston: Houghton Mifflin.

Janis, I. L., & Mann, L. (1977). *Decision making: A psychological analysis of conflict, choice, and commitment.* New York: Free Press.

Jarvenpaa, S. L. (1989). The effect of task demands and graphical format on information processing strategies. *Management Science, 35,* 285–303.

Johnson, E. J. (1988). Expertise and decision making under uncertainty: Performance and process. In M. T. H. Chi, R. Glaser, & M. J. Farr (Eds.), *The nature of expertise*. Hillsdale, NJ: Lawrence Erlbaum Associates.

Johnson, E. J., Payne, J. W., & Bettman, J. R. (1988). Information displays and preference reversals. *Organizational behavior and human decision processes, 42*, 1–21.

Johnson, E. J., Schkade, D. A., & Bettman, J. R. (1989). *Monitoring information processing and decisions: The mouselab system*. Unpublished manuscript, Duke University, Durham, NC.

Johnson, P. E., Jamal, K., & Berryman, R. G. (1991). Effects of framing on auditor decisions. *Organizational Behavior and Human Decision Processes, 50*, 75–105.

Kahneman, D., Slovic, P., & Tversky, A. (Eds.). (1982). *Judgment under uncertainty: Heuristics and biases*. New York: Cambridge University Press.

Kahneman, D., & Tversky, A. (1973). On the psychology of prediction. *Psychological Review, 80*, 237–251.

Kahneman, D., & Tversky, A. (1979). Prospect theory: An analysis of decision under risk. *Econometric, 47*, 263–289.

Kelly, C. W., Andriole, S. J., & Daly, J. A. (1981). Computer-based decision analysis: An application to a Middle East evacuation problem. *Jerusalem Journal of International Relations, 5*, 62–84.

Kent, S. (1964). Words of estimated probability. *Studies in Intelligence, 8*, 49–65.

Kim, E., & Lee, J. (1986). An exploratory contingency model of user participation and MIS use. *Information and Management, 11*(2), 87–97.

Kirschenbaum, S. S. (in press). Influence of experience on information gathering strategies. *Journal of Applied Psychology*.

Klayman, J., & Ha, Y. (1987). Confirmation, disconfirmation, and information in hypothesis testing. *Psychological Review, 94*, 211–228.

Klein, G. A. (1989). Recognition-primed decisions. In W. B. Rouse (Ed.), *Advances in man-machine systems research: Vol. 5* (pp. 47–92). Greenwich, CT: JAI Press.

Klein, G. A. (in press). A recognition-primed decision (RPD) model of rapid decision making. In G. A. Klein, J. Orasanu, R. Calderwood, & C. E. Zsambok (Eds.), *Decision making in action: Models and methods*. Norwood, NJ: Ablex.

Klein, G. A., Calderwood, R., & MacGregor, D. (1989). Critical decision method for eliciting knowledge. *IEEE Transactions on Systems, Man, and Cybernetics, 19*, 462–472.

Kramer, A., Wickens, C., & Donchin, E. (1983). An analysis of the processing requirements of a complex perceptual-motor task. *Human Factors, 25*, 597–621.

Lane, D. M. (1982). Limited capacity, attention allocation and productivity. In W. C. Howell & E. A. Fleishman (Eds.), *Human performance and productivity: Information processing and decision making*. Hillsdale, NJ: Lawrence Erlbaum Associates.

Ledgard, H., Singer, A., & Whiteside, J. (1981). *Directions in human factors for interactive systems*. New York: Springer-Verlag.

Lictenstein, S., Fischhoff, B., & Phillips, L. D. (1982). Calibration of probabilities: The state of the art to 1980. In D. Kahneman, P. Slovic & A. Tversky (Eds.), *Judgment under uncertainty: Heuristics and biases*. New York: Cambridge University Press.

Lictenstein, S., & Slovic, P. (1973). Response-induced reversals of preferences in gambling: An extended replication in Las Vegas. *Journal of Experimental Psychology, 101*, 16–20.

Ligon, J. M. (1989). *Causal modeling for requirements engineering*. Fairfax, VA: George Mason University.

Loftus, E. F. (1979). *Eyewitness testimony*. Cambridge, MA: Harvard University Press.

Lusk, C. M., & Hammond, K. R. (1991). Judgment in a dynamic task: Microburst forecasting. *Journal of Behavioral Decision Making, 4*, 55–73.

March, J. G. (1978). Bounded rationality, ambiguity, and the engineering of choice. *Bell Journal of Economics, 9*, 587–608.

Martin, J. (1991). *Rapid application development*. New York: Macmillan.

Mauch, J. E., & Birch, J. W. (1989). *Guide to the successful thesis and dissertation*. New York: Marcel Dekker.

McNeil, B. J., Pauker, S. G., Sox, H. C., Jr., & Tversky, A. (1982). On the elicitation of preferences for alternative therapies. *New England Journal of Medicine, 306*, 1259–1262.

Meister, D. (1976). *Behavioral foundations of system development*. New York: John Wiley & Sons.

Meyer, D. E. (1970). On the representation and retrieval of stored semantic information. *Cognitive Psychology, 1*, 242–297.

Miller, G. A. (1956). The magical number seven, plus or minus two: Some limits on our capacity for processing information. *The Psychological Review, 63*, 81–97.

Miller, M. J., Brehmer, B., & Hammond, K. R. (1970). Communication and conflict reduction: A cross-cultural study. *International Journal of Psychology, 5*, 75–87.

Miller, R. R. (1991). *The interface development engineering methodology*. Moorestown, NJ: Computer Sciences Corporation.

Miller, R. R., & Iavecchia, H. (1993). *I R & D project, AO12*. Moorestown, NJ: Computer Sciences Corporation.

Milter, R. G., & Rohrbaugh, J. (1988). Judgment analysis and decision conferencing for administrative review: A case study of innovative policy making in government. In S. M. Puffer & R. Cardy (Eds.), *Information processing and decision making*. Greenwich, CT: JAI Press.

Minsky, M. (1975). A framework for representing knowledge. In *The psychology of computer vision*. New York: McGraw-Hill.

Mintzberg, H. (1979). *The structuring of organizations*. Englewood Cliffs, NJ: Prentice-Hall.

Monty, R. A. (1967). Stimulus characteristics and spatial encoding in sequential short-term memory. *Journal of Psychology, 65*, 109–116.

Moore, P. G. (1977). The manager struggles with uncertainty. *Journal of the Royal Statistical Society, Series A (General), 140*, 129–148.

Moray, N. (1977). *Workload measurement*. New York: Plenum Press.

Mumpower, J. L. (1988). The analysis of the judgmental components of negotiation and a proposed judgmentally-oriented approach to negotiation. In B. Brehmer & C. R. B. Joyce (Eds.), *Human judgment: The SJT view* (pp. 465–502). Amsterdam: North-Holland.

Murphy, E. D., & Mitchell, C. M. (1986). Cognitive attributes: Implications for display design in supervisory control systems. *International Journal Man-Machine Studies, 25*, 411–438.

Newell, A. (1981). *The knowledge level* (Rep. No. CMU-CS-81-131). Pittsburgh, PA: Carnegie Mellon University.

Newell, A., & Simon, H. A. (1972). *Human problem solving*. Englewood Cliffs, NJ: Prentice-Hall.

Newton, J. R. (1965). Judgment and feedback in a quasi-clinical situation. *Journal of Personality and Social Psychology, 1*, 336–342.

Nisbett, R., & Ross, L. (1980). *Human inference: Strategies and shortcomings of social judgment*. Englewood Cliffs, NJ: Prentice-Hall.

Norman, D. A. (1980). Discussion: Teaching, learning and the representations of knowledge. In R. E. Snow, P. A. Federico, & W. E. Montague (Eds.), *Aptitude, learning, and instruction, Vol. 2: Cognitive process analyses of learning and problem solving*. Hillsdale, NJ: Lawrence Erlbaum Associates.

Norman, D. A. (1982, August). Human cognition. In L. S. Abbott (Ed.), *Proceedings of Workshop on Cognitive Modeling of Nuclear Plant Control Room Operators*. (NUREG/CR-3114; ORNL/TM-8614).

Norman, D. A. (1986). Cognitive engineering. In D. A. Norman & S. W. Draper (Eds.), *User centered system design*. Hillsdale, NJ: Lawrence Erlbaum Associates.

Norman, D. A., & Bobrow, D. G. (1986). Descriptions: An intermediate stage in memory retrieval. *Cognitive Psychology, 11*, 107–123.

Norman, D. A., & Draper, S. W. (Eds.). (1986). *User-centered system design.* Hillsdale, NJ: Lawrence Erlbaum Associates.

Nisbett, R. E., & Ross, L. (1980). *Human inference: Strategies and shortcomings of social judgment.* Englewood Cliffs, NJ: Prentice-Hall.

Nisbett, R. E., & Wilson, T. (1977). Telling more than we can know: Verbal reports on mental processes. *Psychological Review, 84,* 231–259.

Noble, D. (1989). Schema-based knowledge elicitation for planning and situation assessment aids. *IEEE Transactions on Systems, Man, and Cybernetics, SMC-19,* 473–482.

North, R. L. (1988, January–February). Neurocomputing: Its impact on the future of defense systems. *Defense Computing.*

Northcraft, G. B., & Neale, M. A. (1987). Experts, amateurs, and real estate: An anchoring and adjustment perspective on property pricing decisions. *Organizational Behavior and Human Decision Processes, 39,* 84–97.

Olson, J. R., & Nilsen, E. (1987). Analysis of the cognition involved in spreadsheet software interaction. *Human-Computer Interaction, 3,* 309–349.

Orasanu, J., & Connolly, T. (1991). The reinvention of decision making. In G. A. Klein, J. Orasanu, R. Calderwood, & C. E. Zsambok (Eds.), *Decision making in action: Models and methods.* Norwood, NJ: Ablex Publishing.

Patel, V. L., & Groen, G. J. (1991). The general and specific nature of medical expertise. A critical look. In K. A. Ericsson & J. Smith (Eds.), *Toward a general theory of expertise: Prospects and limits* (pp. 93–125). New York: Cambridge University Press.

Payne, J. W. (1976). Task complexity & contingent processing in decision making: An information search & protocol analysis. *Organizational Behavior & Human Performance, 16,* 366–387.

Payne, J. W. (1982). Contingent decision behavior. *Psychological Bulletin, 92,* 382–402.

Payne, J. W., Bettman, J. R., & Johnson, E. J. (1987). *Adaptive strategy in decision making.* (Tech. Rep.). Durham, NC: Duke University.

Payne, J. W., Laughhunn, D. J., & Crum, R. (1980). Translation of gambles and aspiration effects in risky choice behavior. *Management Science, 26,* 1039–1060.

Pennington, N. (1988). Explanation-based decision making: Effects of memory structure on judgment. *Journal of Experimental Psychology: Learning, Memory & Cognition, 14,* 521–533.

Pennington, N., & Hastie, R. (1986). Evidence evaluation in complex decision making. *Journal of Personality & Social Psychology, 51,* 242–258.

Peterson, C. R., Hammond, K. R., & Summers, D. A. (1965). Multiple probability learning with multiple shifting weights of cues. *American Journal of Psychology, 74,* 660–663.

Pinker, S. (Ed.). (1985). *Visual cognition.* Cambridge, MA: MIT/Bradford Books.

Posner, M. I. (1973). *Cognition: An introduction.* Glenview, IL: Scott, Foresman, & Company.

Potosnak, K. (1989, September). Mental models: Helping users understand software. *Human Factors,* 85–88.

Poulton, E. C. (1968). Searching for letters or closed shapes in simulated electronic displays. *Journal of Applied Psychology, 52,* 348–356.

Ragland, C. (1989, Spring). Hypermedia: The multiple message. *MacTech Quarterly,* 123–129.

Ramsey, H. R., & Atwood, M. E. (1979). *Human factors in computer systems: A review of the literature.* Englewood, CO: Science Applications.

Rasmussen, J. (1985). The role of hierarchical knowledge representation in decision making and system management. *IEEE Transactions on Systems, Man, and Cybernetics, SMC-15,* 234–243.

Rasmussen, R. (1986). *Information processing and human-machine interaction: An approach to cognitive engineering.* New York: North-Holland.

Raymond, L. (1985). Organizational characteristics and MIS success in the context of small business. *MIS Quarterly, 9,* 37–52.

Raymond, L. (1987). Validating and applying user satisfaction as a measure of MIS success in small organizations. *Information and Management, 12,* 23–26.

Reid, G. B., & Nygren, T. E. (1988). The subjective workload assessment technique: A scaling procedure for measuring mental workload. In P. A. Hancock & N. Meshkati (Eds.), *Human mental workload* (pp. 185–218). Amsterdam: North-Holland.

Rogers, Y., Rutherford, A., & Bibby, P. A. (1992). *Models in the mind: Theory, perspective and application.* New York: Academic Press.

Rohrbaugh, J. (1986). *Policy PC: Software for judgment analysis.* Albany, NY: Executive Decision Services.

Rohrbaugh, J. (1988). Cognitive conflict tasks and small group processes. In B. Brehmer & C. R. B. Joyce (Eds.), *Human judgment: The SJT view* (pp. 199–226). Amsterdam: North-Holland.

Rouse, W. B. (1983). Models of human problem solving: Detection, diagnosis, and compensation for system failures. *Automatica, 19,* 613–626.

Rumelhart, D. (1975). Notes on a schema for stories. In D. Bobrow & A. Collins (Eds.), *Representation and understanding.* New York: Academic Press.

Rumelhart, D. E., & Ortony, A. (1977). The representation of knowledge in memory. In R. C. Anderson, R. J. Spiro, & W. E. Montague (Eds.), *Schooling and the acquisition of knowledge.* Hillsdale, NJ: Lawrence Erlbaum Associates.

Rushinek, A., & Rushinek, S. (1986). What makes business users happy? *Communications of the ACM, 29*(7), 455–461.

Russo, J. E. (1977). The value of unit price information. *Journal of Marketing, 14,* 193–201.

Russo, J. E., & Schoemaker, P. J. H. (1989). *Decision traps: The ten barriers to brilliant decision-making & how to overcome them.* New York: Doubleday.

Saaty, T. L. (1980). *The analytic hierarchy process.* New York: McGraw-Hill.

Sage, A. P. (1986). Collective enquiry. In M. Singh (Editor in Chief), *Systems and control encyclopedia.* Oxford, England: Pergamon Press.

Sage, A. P. (1991). *Decision support systems engineering.* New York: Wiley-Interscience.

Sage, A. P. (1992). *Systems engineering.* New York: John Wiley & Sons.

Sage, A. P., & Rouse, W. B. (1986). Aiding the decision-maker through the knowledge-based sciences. *IEEE Transactions on Systems, Man and Cybernetics, SMC-16,* 4, 123–132.

Satterthwaite, F. E. (1946). An approximate distribution of estimates of variance components. *Biometrics Bulletin, 2,* 110–114.

Schank, R. C., & Abelson, R. P. (1977). *Scripts, plans, goals, and understanding.* Hillsdale, NJ: Lawrence Erlbaum Associates.

Schiflett, S. G., Strome, D. R., Eddy, D. R., & Dalrymple, M. (1990). Aircrew evaluation sustained operations performance (AESOP): a triservice facility for technology transition. USAF School of Aerospace Medicine. Brooks Air Force Base, TX.

Schwab, D. (1980). Construct validity in organizational behavior. In B. Staw & L. Cummings (Eds.), *Research in organizational behavior.* Greenwich, CT: JAI Press.

Schweiger, D. M., Anderson, C. R., & Locke, E. A. (1985). Complex decision making: A longitudinal study of process and performance. *Organizational Behavior & Human Decision Processes, 36,* 245–272.

Selcon, S. J., & Taylor, R. M. (1991). Workload or situational awareness?: TLX vs. SART for aerospace systems design evaluation. *Proceedings of the Human Factors Society 35th Annual Meeting* (pp. 62–66). Santa Monica, CA: Human Factors Society.

Serfaty, D., Entin, E., & Tenney, R. (1989). Planning with uncertain and conflicting information. In S. E. Johnson & A. H. Levis (Eds.), *Science of command and control. Part 2: Coping with complexity* (pp. 91–100). Fairfax, VA: AFCEA International Press.

Shafer, G., & Pearl, J. (Eds.). (1990). *Readings in uncertain reasoning.* Palo Alto, CA: Morgan Kaufman.

Simon, H. A. (1955). A behavioral model of rational choice. *Quarterly Journal of Economics, 69,* 99–118.

Simon, H. A. (1960). *The new science of management decisions.* New York: Harper & Row.

Simon, H. A. (1979). Rational decision making in business organizations. *American Economic Review, 69,* 493–513.

Simon, H. A., & Hayes, J. R. (1976). The understanding process: Problem isomorphs. *Cognitive Psychology, 8,* 165–190.

Sheridan, T. (1980). Human error in nuclear power plants. *Technology Review, 82*(4), 22–33.

Shiffrin, R. M. (1976). Capacity limitations in information processing, attention, and memory. In W. K. Estes (Ed.), *Handbook of learning and cognitive processes, Vol. 4, attention and memory.* Hillsdale, NJ: Lawrence Erlbaum Associates.

Shneiderman, B. (1980). *Software psychology.* Cambridge, MA: Winthrop Publishers.

Shneiderman, B. (1983). Direct manipulation: A step beyond programming languages. *Computer, 16,* 57–69.

Shneiderman, B. (1987). *Designing the user interface: Strategies for effective human–computer interaction.* Reading, MA: Addison-Wesley.

Shneiderman, B. (1992). *Designing the user interface: Strategies for effective human-computer interaction* (2nd ed.). Cambridge, MA: Winthrop Publishers.

Sidorsky, R. C. (1982). *Color coding in tactical display-help or hindrance.* Army Research Institute.

Silverstein, L. D. (1987). Human factors for color display systems: Concepts, methods, and research. In H. J. Durrett (Ed.), *Color and the computer.* FL: Academic Press.

Slovic, P., Fischhoff, B., & Lichenstein, S. (1976). Cognitive processes and societal risk taking. In J. S. Carroll & J. W. Payne (Eds.), *Cognition and social behavior.* Hillsdale, NJ: Lawrence Erlbaum Associates.

Slovic, P., Fischhoff, B., & Lichenstein, S. (1977). Behavioral decision theory. *Annual Review of Psychology, 28,* 1–39.

Slovic, P., Fleissner, D., & Bauman, W. S. (1972). Analyzing the use of information in investment decision making: A methodological proposal. *Journal of Business, 45,* 283–301.

Smith, S. L., & Goodwin, N. C. (1971). Blink coding for information displays. *Human Factors, 13,* 283–290.

Smith, S. L., & Goodwin, N. C. (1972). Another look at blinking displays. *Human Factors, 14,* 345–47.

Smith, S. L., & Mosier, D. (1984). *Design guidelines for the user interface to computer-based information systems.* Bedford, MA: The Mitre Corporation.

Smith, S. L., & Mosier, J. N. (1986). *Guidelines for user interface software.* (ESD-TR-86-278). Bedford, MA: Mitre Corporation.

Speroff, T., Connors, A. F., & Dawson, N. V. (1989). Lens model analysis of hemodynamic status in the critically ill. *Medical Decision Making, 9,* 243–252.

Spetzler, C. S., & Stael von Holstein, C. A. S. (1975). Probability encoding in decision analysis. *Management Science, 22,* 340–358.

Srinivasan, A. (1985). Alternative measures of system effectiveness: Associations and implications. *MIS Quarterly, 9,* 243–253.

Sterman, J. D. (1989). Modeling managerial behavior: Misperceptions of feedback in a dynamic decision making experiment. *Management Science, 35,* 321–339.

Stillings, N. A., Feinstein, M. H., Garfield, J. L., Rissland, E. L., Rosenbaum, D. A., Weisler, S. E., & Baker-Ward, L. (1987). *Cognitive Science: An introduction.* Cambridge, MA: MIT Press.

Summers, D. A. (1969). Adaptation to changes in multiple probability tasks. *American Journal of Psychology, 82,* 235–240.

Summers, D. A., Taliaferro, J. D., & Fletcher, D. (1970). Subjective versus objective description of judgment policy. *Psychonomic Science, 18,* 249–250.

Tait, P., & Vessey, I. (1988). The effect of user involvement on system success: A contingency approach. *MIS Quarterly, 12,* 91–108.

Taylor, R. M. (1989). *Situational awareness rating technique (SART): The development of a tool for aircrew systems design.* (AGARD-CP-478) Agard Conference on Situational Awareness in Aerospace Operations.

Tolcott, M. A., Marvin, F. F., & Bresnick, T. A. (1989, June). The confirmation bias in evolving decisions. *Proceedings of the Third Annual Symposium on Command and Control Research.* Washington, DC.

Tolcott, M. A., Marvin, F. F., & Lehner, P. E. (1989a). Expert decision making in evolving situations. *IEEE Transactions on Systems, Man, and Cybernetics, 19,* 606–615.

Tolcott, M. A., Marvin, F. F., & Lehner, P. E. (1989b). Decision making in evolving situations. *IEEE Transactions on Systems, Man, and Cybernetics, 19,* 606–615.

Tolman, E. C., & Brunswik, E. (1935). The organism and the causal texture of the environment. *Psychological Review, 42,* 43–77.

Turner, A. A. (1990). *Mental models and user-centered design.* (ARI Research Note 90-82). Alexandria, VA: Army Research Institute.

Tversky, A., & Kahneman, D. (1971). The belief in the 'Law of Small Numbers.' *Psychological Bulletin, 76,* 105–110.

Tversky, A., & Kahneman, D. (1973). Availability: A heuristic for judging frequency and probability. *Cognitive Psychology, 5,* 207–232.

Tversky, A., & Kahneman, D. (1980). Causal schemas in judgment under uncertainty. In M. Fishbein (Ed.), *Progress in social psychology.* Hillsdale, NJ: Lawrence Erlbaum Associates.

Tversky, A., & Kahneman, D. (1981). The framing of decisions and the psychology of choice. *Science, 211,* 453–458.

Tversky, A., & Kahneman, D. (1982). Judgments of and by representativeness. In D. Kahneman, P. Slovic, & A. Tversky (Eds.), *Judgment under uncertainty: Heuristics and biases.* Cambridge: Cambridge University Press.

Tversky, A., & Kahneman, D. (1983). Extensional versus intuitive reasoning: The conjunction fallacy in probability judgment. *Psychological Review, 90,* 293–315.

Underwood, G. (1976). *Attention and memory.* New York: Pergamon Press.

Vidulich, M. A. (1989). The use of judgment matrices in subjective workload assessment: The subjective workload dominance technique. *Proceedings of the Human Factors Society 33rd Annual Meeting* (pp. 1406–1410). Santa Monica, CA: Human Factors Society.

Vidulich, M. A., & Tsang, P. S. (1987). Absolute magnitude estimation and relative judgment approaches to subjective workload assessment. In *Proceedings of the Human Factors Society 31st Annual Meeting* (pp. 1057–1061). Santa Monica, CA: Human Factors Society.

Vidulich, M. A., Ward, G. F., & Schueren, J. (1991). Using the subjective workload dominance technique for projective workload assessment. *Human Factors, 33,* 677–691.

Wason, P. C. (1960). On the failure to eliminate hypotheses in a conceptual task. *Quarterly Journal of Experimental Psychology, 12,* 129–140.

Watson, S., & Buede, D. M. (1987). *Decision synthesis: The principles and practice of decision analysis.* Cambridge, England: Cambridge University Press.

Weiss, J. J., & Zwahlen, G. W. (1982). The structured decision conference: A case study. *Hospital and Health Services Administration, 27,* 90–105.

Welford, A. T. (1978). Mental workload as a function of demand, capacity, strategy and skill. *Ergonomics, 21,* 151–167.

Wickens, C. D. (1984). *Engineering psychology and human performance.* Glenview, IL: Scott, Foresman, & Company.

Wickens, C. D., & Kessel, C. (1981). Failure detection in dynamic systems. In J. Rasmussen & W. B. Rouse (Eds.), *Human detection and diagnosis of system failures.* New York: Plenum Press.

Wickens, C., Sandry, D., & Vidulich, M. (1983). Compatibility and resource competition between modalities of input, central processing, and output. *Human Factors, 25,* 227–248.

Williams, M. D., & Hollan, J. D. (1981). The process of retrieval from very long-term memory. *Cognitive Science, 5*, 87–119.

von Winterfeldt, D. (1988). Expert systems and behavioral decision research. *Decision Support Systems, 4*, 461–471.

von Winterfeldt, D., & Edwards, W. (1986). *Decision analysis and behavioral research.* Cambridge, England: Cambridge University Press.

Wolf, S. P., Klein, G. A., & Thordsen, M. L. (1991, May). Decision-centered design requirements. *Proceedings of NAECON.* Dayton, OH.

Wohl, J. G. (1981). Force management decision requirements for Air Force Tactical Command and Control. *IEEE Transactions on Systems, Man, and Cybernetics, SMC-11*, 618–639.

Woods, D. D. (1984). Visual momentum: A concept to improve the cognitive coupling of person and computer. *International Journal of Man-Machine Studies, 21*, 229–244.

Woods, D. D., & Roth, E. M. (1988). Cognitive engineering: Human problem solving with tools. *Human Factors, 30*, 415–430.

Zsambok, C. E., Klein, G. A., Beach, L. R., Kaempf, G. L., Klinger, D. W., Thordsen, M. L., & Wolf, S. P. (1992). *Decision-making strategies in the AEGIS Combat Information Center.* Fairborn, OH: Klein Associates.

Author Index

Z

Subject Index

Printed and bound by CPI Group (UK) Ltd, Croydon, CR0 4YY

17/10/2024

01775688-0013